A Practical Analysis of Sea Breeze Effects on Coastal Areas

(with Implications Associated with Renewable Energy Applications, Environmental Assessments, and Recreational Activities)

Rich Dunk, PhD, CCM

Fulton Books
Meadville, PA

Published by Fulton Books 2024

ISBN 979-8-89221-150-5 (paperback)
ISBN 979-8-89221-151-2 (digital)

This writing is dedicated to my immediate family for their love and encouragement—my wife and soulmate, Beryl; our kids, Rick and Becky; their spouses, Steph Dunk and Kiel Eckhardt; and our grandkids—Mikayla, Richie Dunk, Hayden, and Scarlett Eckhardt.

Memoriam

A special tribute is paid to my mentors at USMR/ AMAX, Inc., Gene Deutch, VP of Environmental/ Regulatory Affairs, and Milt Hauser, Director of Environmental Services/Company Metallurgist, for their expert advice and support for both my educational and career endeavors.

Contents

List of Figures and Tables

Tables

Preface

A consensus of people thinks the sea breeze is a local wind that produces refreshing winds from the ocean to shoreline areas during the summer season. However, the sea breeze is a complex three-dimensional thermodynamic circulation that can propagate significant distances both inland and offshore and can extend vertically to substantial heights above the surface. Therefore, the sea breeze will greatly affect activities associated with coastal areas. The intent of this writing is not to supply mathematical equations, detailed descriptions that define the physical processes that cause the sea breeze to develop and intensify, or provide in-depth monitoring/modeling validation results. Also, associated extensive data sets that support the relevant equations that define sea breeze dynamics are provided in the referenced studies related to this publication (Grannakopoulou and Nhili 2022; Avissar et al. 1990).

The intent of this writing is to provide realistic practical concepts and applications that are pertinent to where and when the sea breeze is expected to occur, along with its intensity and duration, which will affect areas associated with the coastal environment. Therefore, this publication is primarily being provided to support practical applications needed by students, professional practitioners, organizations, institutions, and interested individuals involved with activities associated with the coastal/offshore environment. These entities should then be able to acquire a "working" knowledge of the effects the sea breeze has on their specific projects to ensure that these activities are realistic and viable.

In this writing, there are implications and suggestions for further studies that would enhance the understanding of the complex physics and dynamics inherent in the sea breeze circulation. The current and proposed studies should provide the information that is necessary for the efficient planning and cost-effective operational endeavors involved with several applications associated with coastal/offshore project development and subsequent implementation.

The presented conceptual information was primarily derived from studies and subsequent data acquired from a comprehensive ten-year applied research program funded by the NJ Board of Public Utilities (NJBPU) Clean Energy Program (CEP) and conducted by the Rutgers Center for Ocean Observation Leadership, RU-COOL (Dunk et al. 2015).

Although the innovative oceanic and atmospheric modeling programs described herein are intended to enhance the understanding of NJ's coastal/offshore wind resource characteristics, the resultant concepts can be applied to other US mid-eastern coastal areas (e.g., approximately from southern NY to northern NC) with similar topographical characteristics and shoreline configurations, which affect the dynamics of the sea breeze circulation.

1. Introduction

1.1 Sea Breeze Description

The sea breeze is an area-specific thermodynamic process that will have a sustained onshore easterly flow that becomes dominant over a weaker synoptic flow coming from the opposite westerly sector. The sea breeze will propagate inland until the sea breeze and opposing synoptic flow reach equilibrium. At this location (i.e., the sea breeze "front"), the flow will become vertical because of both free and forced convection until the synoptic flow becomes dominant, causing a return flow at a higher altitude back toward the sea. The sea breeze can occur on any day of the year, depending on whether the land, sea, and air parameters are conducive to its development. However, the sea breeze along the US east coast is most prevalent during the mid-spring and summer seasons. This diurnal event usually develops from mid-morning to early afternoon and can be sustained well into the evening hours.

At some location, from the coast to an area offshore, cooling (reduction in energy) will occur, causing the flow to move downward (i.e., subsidence). The flow will then advect near the sea surface toward shore and advance inland, creating the "typical" sea breeze cell that will exhibit the following characteristics: onshore/offshore wind vectors along with varying turbulence, wind shear, and stability properties. Consequently, as a result of the significant spatial and temporal variability inherent in sea breeze circulations, these localized wind resource flow perturbations will have a direct impact on offshore and coastal development and implementation activities.

Previous and current analyses of various sea breeze occurrences indicate that sea breeze variability becomes one of the most prominent aspects in the decision process for both design and application activities associated with coastal/offshore areas. Furthermore, high-resolution modeling results have identified four "types" of sea breeze circulations that can occur along coastal locations (Bowers 2004). These sea breeze types can be defined according to their dimensions, durations, and flow characteristics. For example, the standard sea breeze circulation (cell) has the largest horizontal and vertical dimensions when compared to the other types of sea breeze circulations described in the forthcoming discussion. Sea breeze types are dependent on the intensity of the horizontal sea/land temperature gradient and prevailing atmospheric synoptic conditions. Wind and temperature vertical profiles within and external of each sea breeze "type" will have different physical properties, which will assist in identifying and classifying each sea breeze occurrence.

1.2 Sea Breeze Classification (Type)

1.2.1 Standard ("Pure") Sea Breeze

Standard sea breeze circulations are caused by relatively strong temperature gradients from the sea (cool) to the land (warm), creating an intense temperature gradient and producing a strong flow from the sea toward inland areas. The standard sea breeze circulation (cell) is characterized by having the largest horizontal and vertical dimensions when compared to the other types of sea breeze occurrences. The following diagram describes the structure and physical properties of the standard (pure) sea breeze circulation:

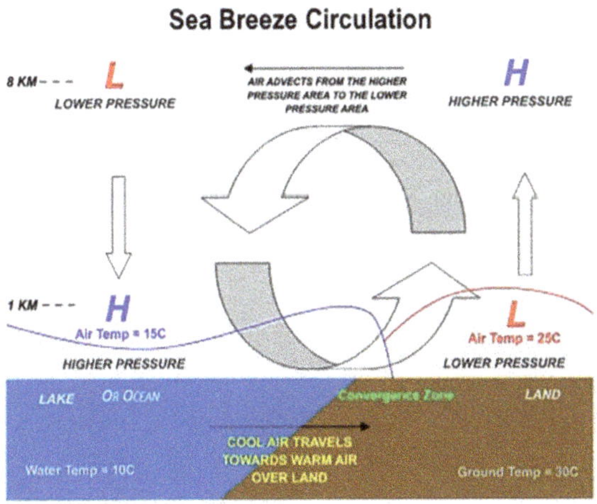

Fig 1. Standard (pure) sea breeze circulation and associated physical parameters

The following visible satellite images with overlays of flow vectors show the progressive subsidence effect on cloud cover and wind intensities, which occurred over the area that coincides with the offshore component of the sea breeze. As a result of the subsiding air over the ocean, cloud cover dissipates and wind speeds are reduced.

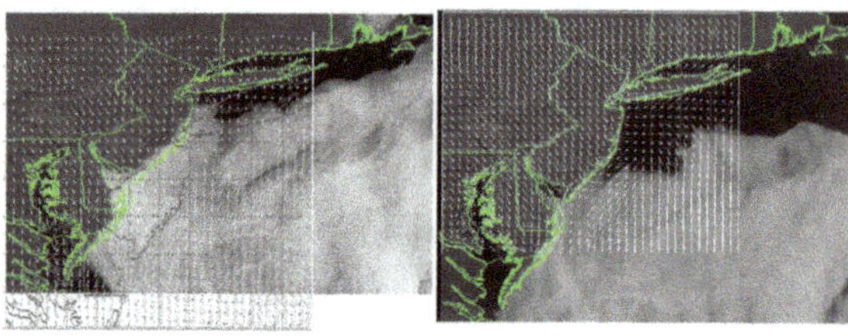

Fig 2. Visible satellite depiction of cloud cover over NJ and Long Island, NY offshore areas, along with an overlay of modeled wind vectors for a concurrent sea breeze occurrence. The images show that as the sea breeze develops, cloud cover progressively dissipates, and wind intensities become reduced over the area effected by the offshore component of the sea breeze.

1.2.2 Side Door ("Corkscrew") Sea Breeze

When relatively strong southwesterly synoptic winds are blowing nearly parallel to the coast and occur concurrently with a temperature gradient from the sea (cold/cool) to the land (warm), a sea breeze will generally develop as being dependent on the intensity of the synoptic flow, along with the magnitude of the temperature gradient between the sea and land. Consequently, the resulting onshore sea breeze flow will be at an acute angle in relation to the prevailing synoptic flow trajectories, creating a "corkscrew" effect in the coastal wind pattern. As the sea breeze intensifies, the angle of the sea breeze wind vectors could approach 90° (i.e., the sea breeze wind vectors could become nearly perpendicular to the synoptic flow streamlines). The sea breeze winds will therefore be coming from the *side* of the synoptic flow, advecting along the coast, and thus the name "Side Door" is given to this sea breeze type.

Side Door sea breeze occurrences, associated with coastal upwelling events, have similar characteristics when compared to the standard sea breeze circulation. These factors are, respectively, the result of a large and concentrated sea/land temperature gradient and the "shearing" effect of the synoptic wind trajectories, which can be nearly perpendicular to both the onshore and offshore component intensities that are significantly greater; their durations and both horizontal and vertical extents are dimensionally restricted in the sea breeze cell. The side door sea breeze type can be subdivided into two classifications:

✓ *Upwelling.* When *continuous* coastal upwelling (cold SSTs) occurs along a good portion of the coast, creating a large temperature gradient between the sea (cold)/land (warm) interface and resulting in a strong sea breeze with consistent intense winds. *Intermediate* coastal upwelling (cold SSTs) occurs as individual "pockets" or centers along the coast producing site-specific intense sea breeze events rather than one larger sea breeze occurrence that develops during the *continuous* upwelling case.

- ✓ *Non-Upwelling:* Synoptic winds are not strong enough to produce upwelling, or the thermocline in coastal waters is insignificant with little temperature change between the surface and underlying layers. The resultant sea breeze occurrences are substantially less intense when compared to the *upwelling* case.

1.2.3 Backdoor Sea Breeze

When synoptic flows are blowing from the northeasterly sector, a "virtual" sea breeze, similar to the southeasterly enhancement sea breeze, can occur. The climatology of most eastern US states indicates that northeasterly winds occur primarily during colder weather when offshore SSTs are generally warmer than onshore land temperatures. Therefore, the lack of a significant temperature gradient from water to land will result in no acceleration of winds across the coastline. In most cases, there will actually be a decrease in wind speeds as winds advect from off-shore to inland areas. Consequently, the onshore flow will be considered to be purely synoptic, with no indication of any of the characteristics that define the sea breeze circulation. However, there will be "rare." As a result of strong temperature gradients between cold coastal waters produced by upwelling and warmer offshore waters located adjacent to the upwelling center along with warmer land temperatures, a reasonable hypothesis would be that two sea breeze cells can develop. One cell would develop inland from the coast, and the other cell would develop offshore from the coast. This hypothesis should be investigated to see whether or not two cells form and, if they do, what atmospheric/sea conditions are conducive to dual cell development. The concept of the dual cell development process is depicted in the below schematic:

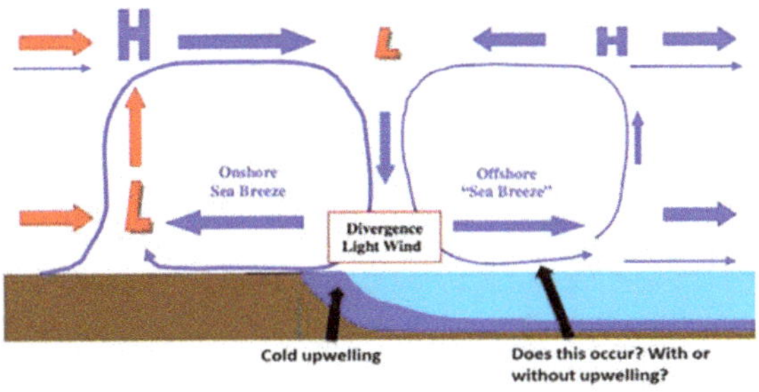

Fig 3. Schematic of the sea breeze cells that can possibly develop during coastal upwelling events

The "corkscrew" effect applies to the "backdoor" sea breeze, except the coastline is to the right of the flow instead of to the left of the flow, as is the case for the southeasterly enhancement sea breeze. Plan views of corkscrew (left) and backdoor (right) sea breeze-generating scenarios, depicting the effect of shore parallel gradient winds on a coastline (green). The black arrows depict the unaltered gradient of wind direction. The red arrows portray frictional effects on the gradient flow at the coastline. Wind flow vectors for, respectively, the "corkscrew" and "backdoor" sea breeze events are shown in the following diagrams:

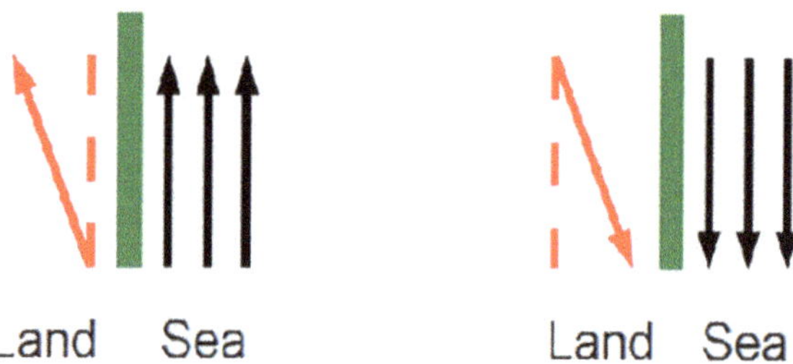

Fig 4. "Corkscrew" and "backdoor" sea breeze flow vectors

1.2.4 Southeasterly Enhanced Sea Breeze

When synoptic flows are blowing from the southeasterly sector, a "virtual" sea breeze can occur. This type of sea breeze occurrence will be dependent on the magnitude of the temperature gradient across the sea/land interface and the strength of the wind speeds associated with the synoptic flow. If the gradient is strong enough, winds will accelerate across the sea/land interface. Therefore, the onshore flow ("sea breeze") will be "enhanced," resulting in the name SE enhancement. However, this "type" of sea breeze appears to occur much less frequently than either the standard or side door sea breeze circulations. The southeasterly enhanced sea breeze will probably have no upper-level return flow, and therefore, it is unlikely that a sea breeze "cell" will be identified. Consequently, according to the definition, this type of sea breeze is not "true" sea breeze circulation. However, several of the characteristics of this specific onshore flow occurrence are similar to those of conventional sea breeze circulations. Therefore, southeasterly enhancement as a "sea breeze" type has a definitive impact on coastal/offshore resources.

1.3 Sea Breeze Variability

The following simulation (Fig. 5) shows the horizontal and vertical cross sections of a *standard* sea breeze occurrence, which indicates that there is significant variability in all aspects of the sea breeze circulation and its impact on the coastal/offshore environment.

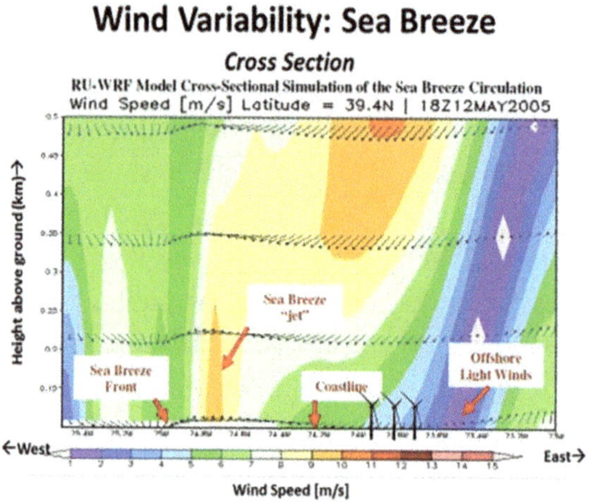

Wind Variability: Sea Breeze

Cross Section

Fig. 5. Cross section of wind variability during a "typical" sea breeze occurrence

Referring to the preceding horizontal and vertical (cross-sectional) simulations for the two sea breeze cases, the following results are indicative of the standard (pure) sea breeze circulation. For the specific sea breeze cases, wind speeds over the offshore locations (designated by blue/purple areas) are < 4 m/s, with the same wind speed being nearly constant from near the sea surface and extending upward through the sea breeze cell.

> ➤ Wind speeds (> 4 to ~6 m/s) and vertical shear become more significant near the sea breeze front that is located onshore in close proximity to the coast. These narrow bands (designated by the red/yellow/orange areas)

are located to the left (west) in both the horizontal and vertical images.

➤ There is an indication of a low-level "jet," with speeds approaching 10 m/s and potentially greater, that are estimated to occur at heights ranging from ~50 m to 200 m above MSL. The "jet" is progressing onshore behind the sea breeze "front," as indicated by the narrow gold-to-red peaks located toward the left within the above simulation.

➤ Previous and current analyses of various sea breeze occurrences indicate that the dynamic sea breeze circulation exhibits substantial temporal and spatial variability within both onshore/offshore areas during early afternoon through late evening hours, which is indicative of the summer season.

A progressive time sequence showing the wind speed variability of a selected sea breeze occurrence, modeled at 10 m over the offshore area, is displayed in the following images:

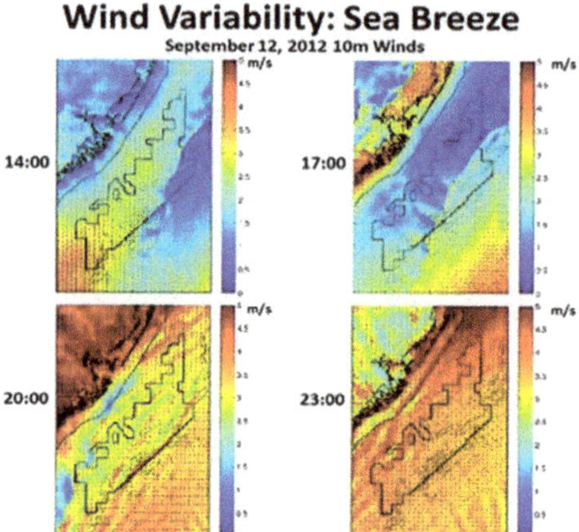

Fig 6. Sea breeze occurrence simulated for an offshore area showing both significant temporal and spatial variability in relatively low wind speeds that ranged from < 0.5 m/s to ~4.5 m/s

The resultant sea breeze characteristics (e.g., wind vectors, turbulence, shear, and dimensions of the circulation system) will change spatially and temporally as the sea breeze intensifies and eventually dissipates during its diurnal cycle. Consequently, along with minimal winds, there could be periods during sea breeze occurrences when wind intensities become significant. Therefore, sea breeze variability and its impact on coastal resources should be considered for any decision-making process or application procedures that are associated with the coastal environment.

2. Sea Breeze Diagnostic and Predictive Procedures

2.1 Monitoring Methods

In-situ and remote-sensing monitoring methods are used for analyzing the sea breeze and for modeling input and verification. Primary support is acquired from atmospheric monitoring systems located near the coast. In-situ monitoring systems, which are shown in the figures on the following page, include a fully instrumented 12-meter meteorological tower (upper figure, pg. 14), a 60-meter meteorological tower, and a 120-meter meteorological tower instrumented at three levels (10m, 60m, and 116m (lower figure, pg. 14). Also, an offshore monitoring buoy (upper figure, pg. 15), equipped with both in-situ and remote LiDAR (light detection and ranging) monitoring systems, along with a SoDAR (sound detection and ranging) remote monitoring system (lower right figure), are used as the primary monitoring systems for analyzing the sea breeze circulation. It should be noted that the anemometry and both the specific SoDAR and LiDAR systems, which can monitor winds at heights up to and greater than 200 m, are accepted by the wind energy industry, and rigorous validation procedures can be used for model data assimilation and validation (USNRC 2007).

12m Meteorological Tower

Fig. 7a. In-situ Meteorological Monitoring Systems used for Sea Breeze Analyses.

Offshore monitoring Buoy

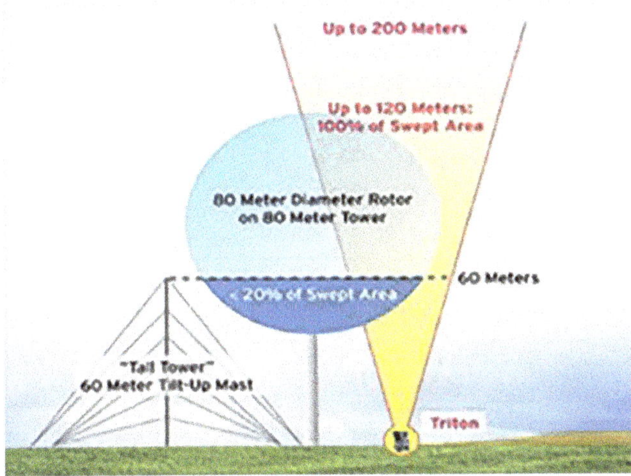

SoDAR System

Fig. 7b. Remote Meteorological Systems used for Sea Breeze Analyses

Representative graphical and digital data outputs from the SoDAR system are depicted in the following figures:

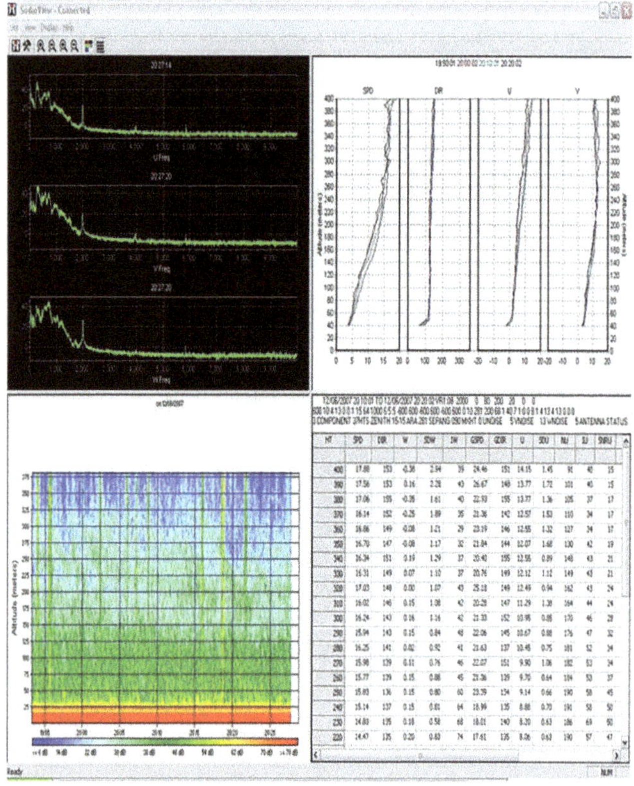

Fig. 8. Example of graphical and digital data acquired from the coastal-located SoDAR system

The tower, buoy, and SoDAR system data used for the sea breeze analyses were supplemented with additional verified data acquired from representative remote-sensing technology that consists of the following systems:

> High-resolution declouded IR satellite surface temperature detection, specifically sea surface temperatures (SSTs) (1st figure)
> Coastal ocean dynamics application radar (CODAR) data acquisition systems (2nd left lower figure) that monitor sea surface waves/currents
> Weather radar (right lower figure) showing the red "backscatter" line, located near the central to southern NJ shoreline, that can detect the sea breeze "front" and its propagation
> Sound detection and ranging (SoDAR) remote sensing of the sea breeze boundary layer dimensions
> Light detection and ranging (LiDAR) system (center bottom figure), which can detect wind and temperature profiles over the area (e.g., coastal or offshore) being analyzed

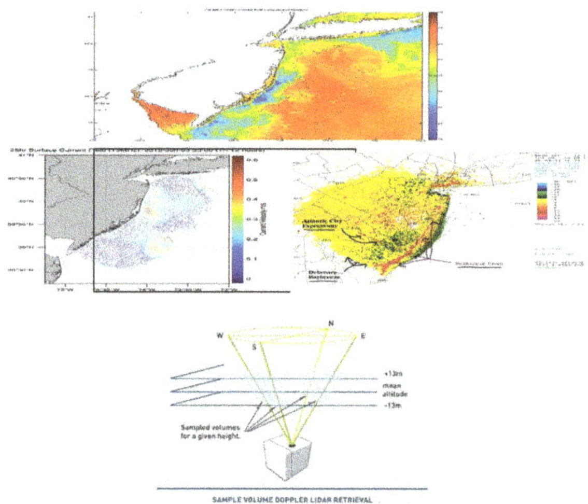

Fig. 9. Supplemental remote-sensing technology used for sea breeze detection and diagnostics

The data relevant to the sea breeze circulation can be acquired from the aforementioned in-situ and remote modeling systems, which are dependent on their location and verification to ensure that the acquired data is representative and accurate. The monitoring systems could be used to ascertain whether a sea breeze has occurred or is occurring. Also, representative and validated monitoring systems can be utilized for model data assimilation and model verification.

The below images depict how the weather radar, when used in conjunction with high-resolution modeling, can effectively "visualize" the sea breeze's development, horizontal extent, and intensity. These images show that the sea breeze develops near the shoreline and propagates well inland (left series of images) during the day, coinciding with increasing winds from offshore areas to inland areas (right series of images). The "darker" blue/green line running from northerly areas to southerly areas "defines" the sea breeze "front" (left series of figures); referring to the right series of figures, the red shading indicates increased convergence/convection, and the blue shading indicates increased divergence/dissipation of wind flows produced by the sea breeze. The following images illustrate "typical" sea breeze development and propagation:

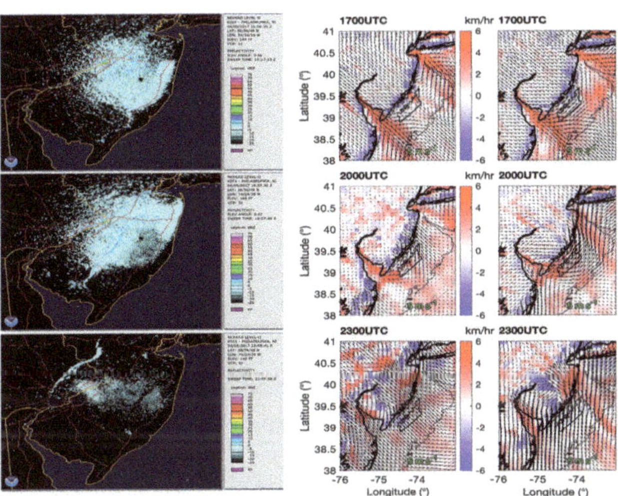

Fig. 10. WX RaDAR and modeling simulations showing the temporal and spatial progression of a "typical" sea breeze event

The sea breeze develops near the coast, intensifies with time, and advects well inland prior to its demise at the end of the day. Coastal wind, turbulence, and atmospheric stability data relevant to sea breeze circulation can be acquired from the aforementioned in-situ and remote modeling systems, which are dependent on their location and verification to ensure that the acquired data is representative and accurate. The monitoring systems could be used to ascertain whether a sea breeze has occurred or is occurring. Also, representative and validated monitoring systems could be utilized for model data assimilation and model verification.

To conduct a comprehensive diagnostic analysis and develop a representative predictive procedure for the sea breeze, a validated mesoscale modeling program should be considered. This suggested modeling effort becomes viable as a result of minimal and nonexistent coastal and offshore meteorological monitoring sites and resultant data. Consequently, a validated atmospheric model that accounts for the physical properties and dynamics of coastal/offshore areas should be utilized for sea breeze analyses. The selected model used for this analysis is the weather research and forecasting (WRF) model that is described in the next section.

2.2 Modeling Methods

2.2.1 Model Selection and Modeling Procedures

Output from the previously described in-situ and remote-sensing monitoring instrumentation is automated to provide real-time data transfer for model input and verification. The latest version of WRF (weather research and forecasting) model (version 3.7) proved to be most applicable for the current coastal/offshore studies (Optis et al. 2020). However, if later versions of WRF or another modeling program will better simulate the coastal/offshore atmospheric parameters, then these modeling alternatives should be evaluated to provide the most representative coastal/offshore analyses based on the "state-of-the-science." The WRF model, which is accepted by most government agencies, the military, academic institutions, and private industry, is used as the basis for the current model simulations.

Since sea surface temperature (SST) is the primary parameter that "dictates" the magnitude of energy exchange between the ocean and overlying atmosphere, temperature differences between the air/sea interface, along with adjacent terrestrial temperatures and resultant vertical/horizontal temperature gradients, will significantly influence the characteristics of both coastal and offshore wind resources. Consequently, representative SSTs are a critical model input parameter for both diagnostic and predictive applications. A new algorithm, which includes visible reflectivity, to differentiate between cloudy and clear conditions associated with IR satellite detection of SSTs has been developed to minimize missing "undetected" data, as experienced with conventional technology.

The innovative high-resolution product, which uses a "dynamic" rather than a "static" ocean commonly used in most mesoscale modeling programs, resolves coastal upwelling centers that are not normally detected with current methods. The "new" SST algorithm has been tested and verified using actual SST measurements acquired from monitoring buoys and autonomous underwater vehicle (AUV) instrumentation (Ref. Rutgers). Furthermore, the new "declouded" product has been automated for model input and appears to enhance both the accuracy and precision of model results. This enhancement to the WRF model was developed and verified by the Rutgers Center for Ocean Observation Leadership (RU-COOL) (Dunk et al. 2015). Therefore, the enhanced model is referred to as the RU-WRF model. This unique capability will substantially enhance the determination of sea breeze development and its potential intensification.

Using area-specific SSTs will define upwelling/non-upwelling centers that are generally not detected by satellite products currently being used by most organizations and institutions. An example of this unique product is provided in the following image:

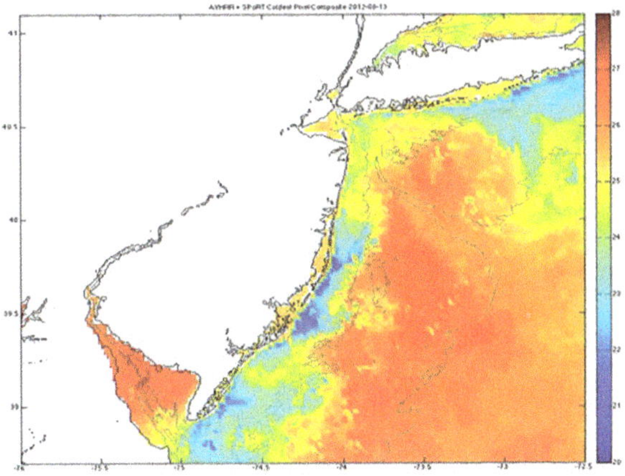

Fig. 11. Satellite SST imagery derived from the RU-COOL Declouded Product showing coastal upwelling [colder temperatures (dark blue) near the central/southern NJ shoreline with warmer temperatures (yellow to red) detected farther offshore]

The RU-WRF model incorporates both atmospheric and oceanic parameters (including the RU-COOL SST product) that will enable an understanding of most physical phenomena that affect the dynamic processes associated with the sea breeze that substantially impact coastal resources. Furthermore, it has been demonstrated that the RU-WRF model provides representative wind and temperature gradients and profiles that are sensitive to the physical processes and geographical configurations of the coastal area being analyzed (Dunk et al. 2015).

The following boundary conditions and the physics used to initialize and "drive" the current model are selected to "best" simulate winds along with other relevant meteorological variables associated with the marine atmospheric boundary layer (MABL), which extends from sea level upward to ~1 km and can extend to ~3 km. The RU-WRF model is currently run at 9 km, 3 km, and 1 km, or at higher horizontal grid resolutions, depending on the physical properties of the specific project being analyzed. Therefore, the mesoscale model can be "nested" to run at microscale resolutions to resolve local wind perturbations (e.g., the sea breeze circulation) within the general flow patterns of the coastal/offshore wind resource. The area modeled at the 9-kilometer resolution is considered the *external* modeling domain, which is used to set initial boundary conditions for the higher-resolution model runs. The area modeled at the 3-kilometer resolution is defined as the internal modeling domain where the modeling "nesting" features will be utilized to resolve "local" events. Vertical levels (altitudes) can be set at increments that will conform to most model applications (e.g., 10 m, 50 m, 100 m, 200 m, 500 m, 1000 m, 2000 m, etc.). The modeling domain is depicted in the following figure:

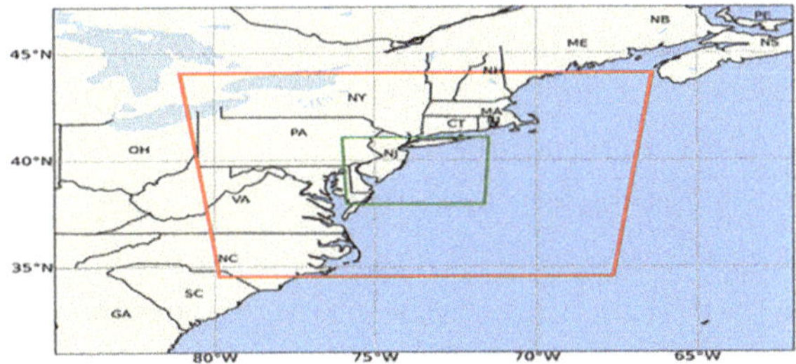

Fig. 12. A "nested" 9- to 3-kilometer domain for the RU-WRF simulations was utilized for coastal areas. The 3-kilometer domain (green boundary) spans from NY to MD, and the 9-kilometer domain (red boundary) spans from ME to NC. More high-resolution modeling with "nesting" features that will simulate wind and temperature at 0.5 to 1 km can be performed for specific sea breeze events.

As implied and demonstrated in the previous discussion, sea breeze circulation can only be effectively analyzed for "real-world" applications through the utilization of a combination of current monitoring and modeling technology. The most unknown factors associated with the sea breeze are the physics, dynamics, and dimensional structure of the offshore component of the sea breeze circulation. As was previously discussed, validated modeling programs in conjunction with representative monitoring data can be utilized to resolve both the offshore and onshore components of the sea breeze circulation. Therefore, by utilizing "state-of-the-science" modeling/monitoring technology, the physics and dynamics of the entire sea breeze circulation can be resolved.

One of the most effective technologies for determining where and when convergence (convection) or divergence (sublimation) occurs within the sea breeze circulation is the application of Lagrangian coherent structures (LCSs), which realistically define both convergence and divergence areas within the sea breeze cell that will determine where and when the sea breeze circulation will have the most effect on coastal areas (Fredj 2016). A description and the utilization of the LCS technology are discussed in the proceeding section.

2.2.2 Lagrangian Coherent Structure (LCS) Model Enhancement

The primary objective of current studies is to provide a better understanding of the dynamics, including the temporal/spatial variability, associated with sea breeze circulation. By combining validated "state-of-the-science" mesoscale modeling (e.g., the RU-WRF model) with LCS routines, we should be able to realistically define the horizontal and vertical cross-sectional dimensions of the coastal/offshore MABL, which encompasses the sea breeze circulation.

LCS and relative dispersion (RD) technology have the potential to objectively clarify and identify the offshore component of the sea breeze in convergent and divergent areas (Seroka 2018). This increased understanding of sea breeze science will lead to improved modeling and prediction of sea breeze circulation. Real-time LCS and RD products in conjunction with forecasts from numerical weather prediction models (e.g., the RU-WRF model) can provide more useful guidance for most entities involved with the energy, environmental, economic, and recreational activities associated with coastal areas. For example, RD forecasts can highlight divergent or repelling structures offshore, which coincide with sinking air and weaker wind speeds, and can also highlight convergent or attracting onshore wind trajectories, which coincide with rising air, a wind shift, high wind shear, and possibly stronger winds that concurrently occur with an intensifying sea breeze circulation.

A description of the Lagrangian methods used to more coherently define the flow field structures associated with the sea breeze circulation is provided in the following discussion. Sea breezes can occur almost daily during the spring, summer, and early fall months. Therefore, the frequency, duration, and intensity of sea breezes in coastal areas and their effects on activities associated with the coastal environment are of paramount importance. The current confidence in sea breeze predictability, especially the detail of each case, is relatively low. Depending on the atmospheric and oceanic conditions, the sea breeze will form at a certain time in late morning or early afternoon, evolve in both time and space in its own unique way, and cease at a time from a few to several hours later. Furthermore, because observations are limited to coastal waters and offshore, the predictability of the sea breeze, especially its offshore component, is not certain.

New and/or improved methods are needed to improve the predictability of and clarify the offshore component of the sea breeze circulation. Unlike the onshore side of the sea breeze, where the inland frontal boundary is clearly defined, the offshore boundary ("front") of the sea breeze circulation is not well defined. Possibly, incorporating the LCS/RD technology into mesoscale modeling programs will enable the offshore structure and dynamics of the sea breeze cell to be identified. Following are Wx RaDAR images that clearly define the onshore sea breeze "front" (red backscatter demarcation line):

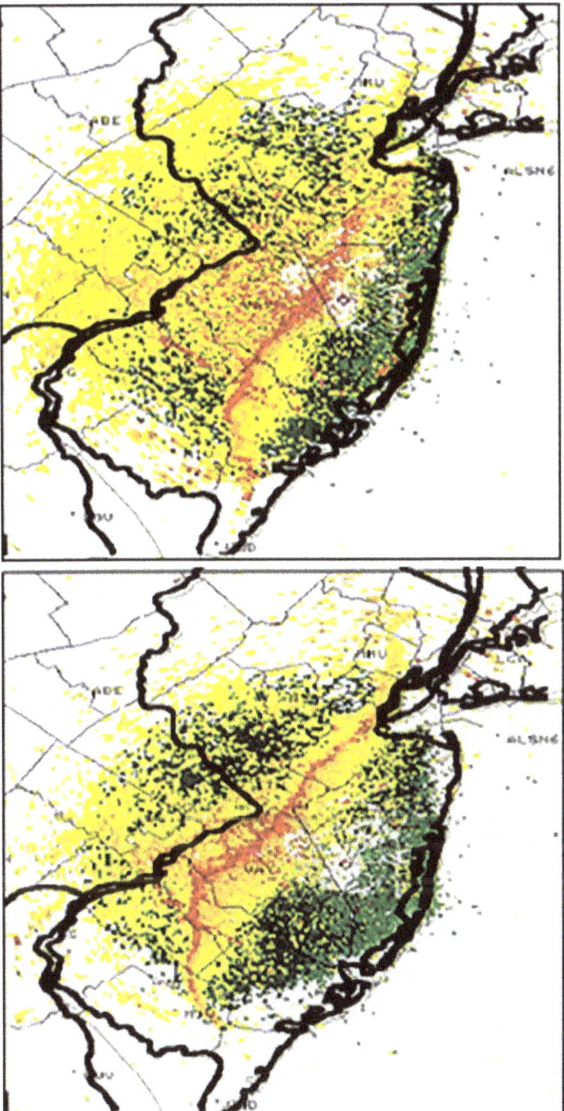

Fig. 13. Sea breeze "front" detected by Wx RaDAR imagery (i.e., backscatter red demarcation line)

Basically, improved sea breeze modeling procedures include the following: 3D LCS trajectories will be transposed on the RU-WRF model wind field vectors to define areas of convergence/divergence along more clearly with areas of associated convection/sublimation to enable a more comprehensive "picture" of the sea breeze circulation.

The following flow diagram illustrates the mesoscale model/LCS 2D trajectory theory for convergence (attracting trajectory "particles") and divergence (repelling trajectory "particles"):

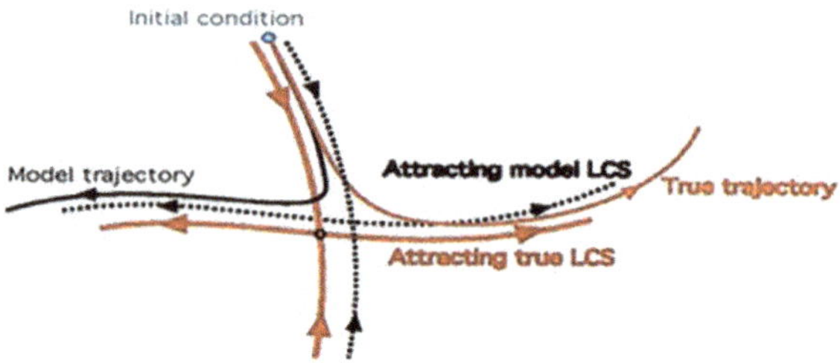

Fig. 14. LCS model trajectories

Simulations show attracting (convergence) and repelling (divergence) trajectories as they change with time (t) and atmospheric pressure (P) relative to the Lagrangian flow field. These hypothetical simulations using the LCS technology are displayed in the following images:

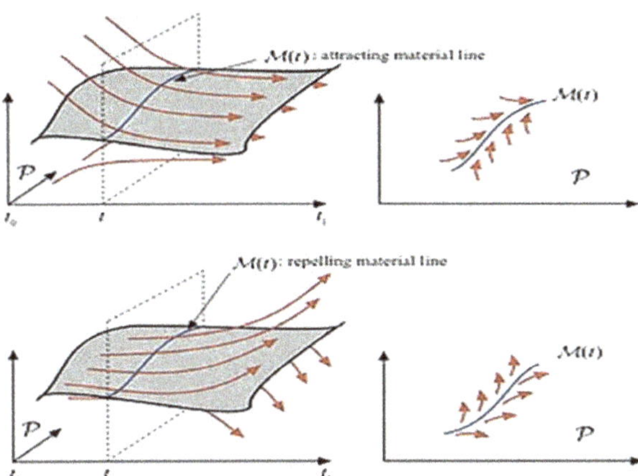

Fig. 15. Attracting (convergence) and repelling (divergence) LCS trajectories relative to changing time and atmospheric pressure

Referring to the previous diagrams, it is apparent that trajectories in a model flow generally show vastly different behavior from trajectories starting with the same initial conditions as the real flow. This is a result of the inevitable accumulation of errors and uncertainties, as well as the sensitive dependence on the initial conditions utilized in any realistic flow model. However, an attracting LCS (such as an unstable array) appears to be remarkably robust with respect to modeling errors and uncertainties. Therefore, LCSs, along with RD technology, are ideal tools for model validation and enhanced predictive routines.

A detailed analysis and description of the LCS technology are presented in the "proof-of-concept" paper provided by *Aqua Wind* (Seroka 2020). The coastal/offshore boundary layer analysis was conducted specifically for parameters associated with the sea breeze circulation, which include:

➢ Vertical temp/wind (Ws, Wd) profiles
➢ Horizontal temp, wind, and pressure gradients at specified grid points for specified coastal areas
➢ Wind shear and turbulence intensity

The preceding parameters would then be used to identify:

➢ temporal and spatial "frontal" formations associated with both the onshore and offshore components, including both convergent/divergent areas of the sea breeze cell and
➢ offshore and onshore boundary layer development/dimensions, which are generally different for each sea breeze event.

To illustrate the variability of the convergence/divergence that occurs over time within the sea breeze circulation, two different time frames—a.m. (top figure) and p.m. (bottom figure) are utilized. The following are example graphics showing the convergence (blue shading) and divergence (red shading) that could occur during a standard ("pure") sea breeze event.

When analyzing the following figures, *the shoreline is designated by the solid black vertical line, the vertical black box-shaped dotted lines define the transition area from sea to land where the sea breeze dynamics begin to be influenced by both offshore and onshore environmental parameters, and the vertical black horizontally shaped dotted line represents the offshore boundary of the sea breeze cell.* Simulations using the LCS technology at 10 m over a "hypothetical" offshore area are displayed in the following graphics:

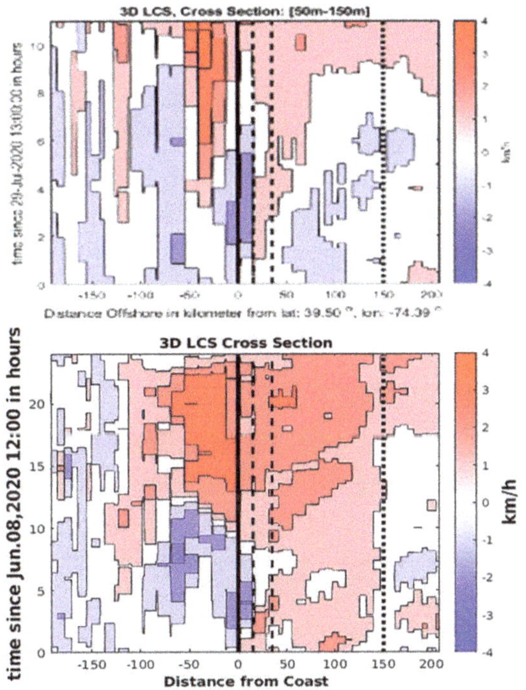

Fig. 16. LCS simulation of a sea breeze temporal/spatial convergence (blue) and divergence (red)

The preceding LC simulations show the variability associated with the convergence and divergence in the northern portion of the sea breeze, along with significant convergence at the shoreline and inland of the central/southern portion of the modeling domain. This is characteristic of a "typical" standard ("pure") sea breeze occurrence. The preceding graphics also indicate that there is significant onshore divergence in the northern portion of the sea breeze, along with significant convergence at the shoreline and inland of the central/southern portion of the modeling domain, which is indicative of the sea breeze circulation dynamics as it propagates inland and extends farther offshore.

Whereas traditional Eulerian fixed frame of reference methods produced from model simulations can be rather "noisy," the Lagrangian "flow-following" frame of reference method has the potential to objectively clarify and identify the offshore component of the sea breeze in convergent and divergent areas. This increased understanding of sea breeze science will lead to improved modeling and prediction of the sea breeze. Therefore, validated monitoring systems, along with approved high-resolution modeling programs, should be conducted to produce more accurate estimates of the land/sea surface conditions and concurrent atmospheric boundary layer characteristics that control sea breeze development and intensification.

Since the sea breeze is a dynamic 3D local circulation, along with validated mesoscale modeling and LCS/RD technology, enhancements should be evaluated and implemented to ensure the accuracy and precision of both diagnostic and predictive results. Possibly, an additional modeling technique should be considered to better characterize sea breeze dynamics, including its spatial dimensions and physical properties (e.g., turbulence, wind shear, and local flow perturbations). The suggested advanced modeling technique is the large eddy simulation (LES) technology, which is a form of computational fluid dynamics (Peng et al. 2022) and should prove to be compatible with mesoscale modeling and LCS/RD routines. The LES enhancement should provide the most representative description of the physical parameters that initiate and control the dynamics of sea breeze circulation. A description of the LES technology is provided in the proceeding discussion.

2.2.3 Large Eddy Simulation (LES) Technology

To account for the variability and controlling physics of the complex flow occurrences associated with the sea breeze circulation, LES technology appears to be the most realistic and representative method for qualifying and quantifying the parameters that develop and control the sea breeze circulation (Grosman 2012). LES modeling of turbulent wakes produced by wind energy turbines and related turbulence parameters is shown in the following images:

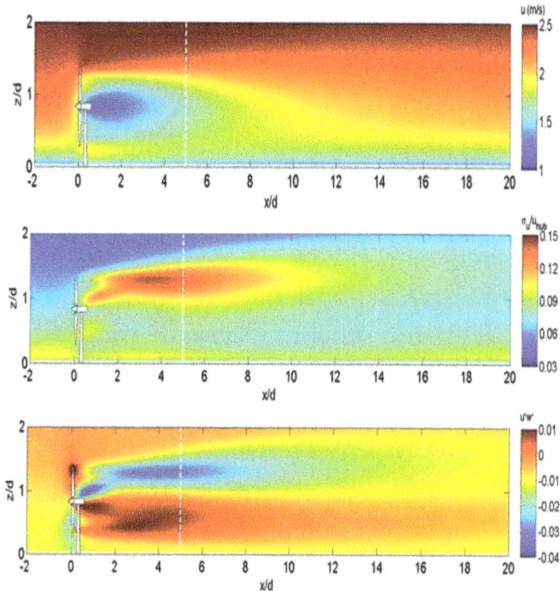

Fig. 17. Simulation results for turbulence parameters obtained using LES with a Lagrangian scale-dependent dynamic model: average velocity (top), turbulence intensity (middle), kinematic shear stress (bottom)

In addition to analyzing turbulence properties, RU-WRF/LCS/LES modeling can be used to provide a detailed area-specific assessment of point-specific wind vectors, wind shear, and flow trajectories that can potentially be encountered during each sea breeze type. The proceeding LES simulation shows a representative 3D wind circulation for a "typical" sea breeze occurrence:

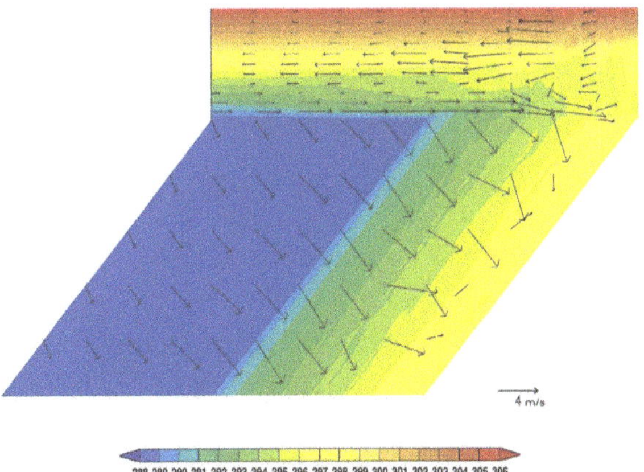

Fig. 18. Perspective view of an LES model sea breeze wind vectors near-surface to ~100 m above MSL

2.2.4 Coupled Atmospheric/Ocean Model

Utilizing the preceding information provided in previous sections regarding area-specific high-resolution monitoring/modeling programs, atmospheric/oceanic events can be analyzed to potentially ensure that a realistic 3D representation of both the coastal/offshore components of the sea breeze circulation is realistic and representative of the atmospheric/oceanic physics that cause sea breeze development and its intensification.

To satisfy the preceding criteria, a "coupled" ocean/atmospheric model can be developed and implemented (Avissar et al. 1990; Dunk et al. 2015). The coupled ocean-atmosphere-wave-sediment transport (COAWST) system framework provided by USGS could be set up and used as the basis for the coupled model. If implemented, the atmospheric model component in this system would consist of the most current version of the RU-WRF model that would include enhanced LCS/RD and LES technologies.

The ocean model component that can be used is the regional ocean modeling system (ROMS), which was developed by and is currently being used in several oceanography projects with a focus on mid-Atlantic coastal/offshore waters. Since the wave component of the coupled model requires extensive computer capabilities, the wave component can be "turned" on or off as needed. A schematic of a suggested "combined" modeling system is provided in the following procedural diagram:

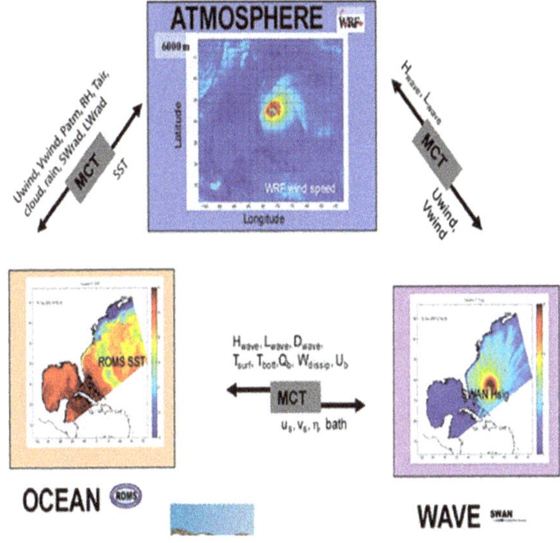

Fig. 19. Coupled modeling system proposed for sea breeze analyses and prediction procedures

Since the RU-WRF model already incorporates the high-resolution SST product into its computational schemes, this could prove to be the only sea parameter needed to account for the influence of the ocean on sea breeze development and intensification. Therefore, a coupled atmosphere/ocean model should be evaluated using a sensitivity study to determine which modeling procedure would be the most practical and representative for both diagnostic and predictive applications associated with sea breeze circulation.

3. Virtual Meteorological Tower Modeling Application

Since coastal/offshore monitoring is very limited or nonexistent, the conceptual utilization of "virtual" meteorological towers (VMTs) should be employed in conjunction with the verified RU-WRF model with the suggested LCS/RD and LES program enhancements to estimate realistic wind and temperature profiles along with turbulence intensity and atmospheric stability at various selected coastal/offshore sites, which would be representative of the MABL associated with the sea breeze dimensions and dynamics. A hypothetical offshore VMT array is shown in the following images:

Fig. 20. Hypothetical offshore VMT locations (Derived VMT parameters would include: wind shear, turbulence intensity, and atmospheric stability.)

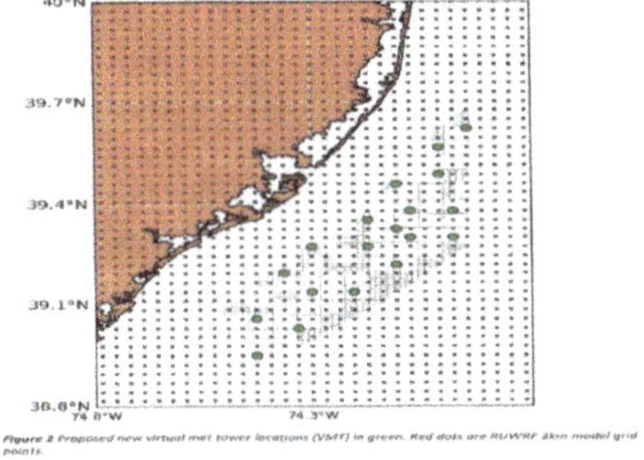

Fig. 21. Hypothetical virtual met tower locations (VMT) in green. Black dots are RU-WRF 3-kilometer model grid points.

The below simulation incorporates all of the suggested modeling methodology, including the current RU-WRF model configuration, LCS + LES technology, and the VMT concept, which imply this is the most effective and efficient methodology to realistically analyze and predict the sea breeze circulation:

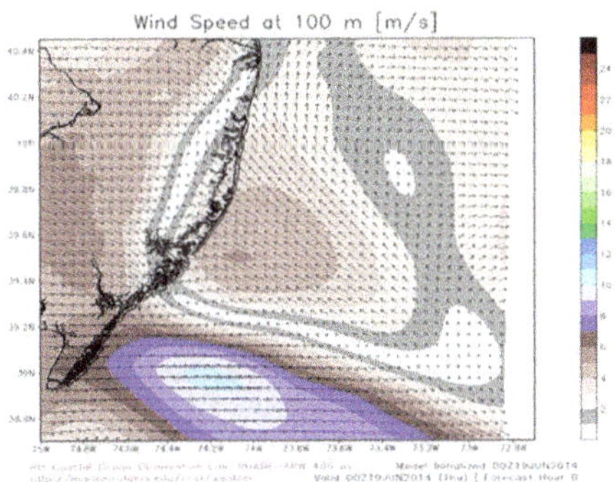

Fig. 22. High-resolution (400-meter horizontal grid spacing) sea breeze simulation at 100 m above MSL, showing significant variability in the offshore wind resource, with wind speeds ranging from <0.2 m/s to ~10 m/s

When analyzing the preceding 100-meter simulation, the following summarized facts were realized regarding the coastal/offshore wind resource:

> Wind speeds over the northern portion of the modeling domain ranged from ~6 m/s nearer the coast and decreased to ~2 m/s farther offshore as a result of intense sea breeze development. This estimated wind speed gradient is contrary to most offshore wind resource assessments, which assume wind speeds increase from the coast to offshore areas. However, this simulation shows wind speeds do increase from ~6 m/s closer to the coast to nearly 11 m/s farther offshore over the southern area of the modeling domain, where the sea breeze did not develop.

> As previously mentioned, the sea breeze circulation is considered a dynamic process with wind flow vectors that constantly change over the diurnal cycle of the circulation.

> Wind speeds over the southern portion of the modeling domain, where the sea breeze did not develop, appear to be more intense when compared to wind speeds estimated for the northern area.

> The turbulence, shear, and physical dimensions of the circulation system will be substantially more variable in the northern portion as the sea breeze intensifies and eventually dissipates during its diurnal cycle.

A detailed analysis of the preceding statements regarding the RU-WRF model applications for sea breeze analyses can be reviewed through this article, "Sea Breeze Sensitivity to WRF Model Configuration; Sea Breeze Turbulence Analysis." It provides an intensive sea breeze modeling study consisting of two parts:

1) Modeling parameters are analyzed to determine which parameters are needed to realistically resolve the wind flow patterns and basic structure of the sea breeze circulation. The parameters studied are model resolution and initial physical conditions. The results of this analysis indicate that the correct initial physical conditions are the most critical model input for resolving the sea breeze circulation. Also, the correct model resolution is important for resolving certain sea breeze circulations that are not well defined spatially or contain complex flow patterns.

2) Turbulence properties associated with sea breeze dynamics are also analyzed and documented.

To enable better sea breeze predictive procedures, the climatology of the sea breeze should be verified to enable a realistic evaluation of expected times, locations, and "type" of sea breeze occurrences. A "short-term" climatological analysis of the sea breeze for most of the mid-east coast is provided in the following discussion.

4. Sea Breeze "Short-Term" Climatology

4.1 Introduction

The primary "trigger" for sea breeze development is the temperature gradient produced by colder SSTs occurring adjacent to warmer inland temperatures. Consequently, cooler, dense air will flow onshore toward warmer, less dense air, producing the sea breeze.

Although the sea breeze can occur in any month of the year, it is most prevalent during the 6-month period of April–September (Robinson 2020). Basically, during the spring through early fall periods, onshore land temperatures are significantly warmer than the SSTs associated with the coastal ocean.

Generally, the converse holds true for the other six months (i.e., October through March). For the preliminary "short-term" sea breeze climatology analysis, the data from the Oyster Creek (Forked River, NJ) met tower, which is redundantly instrumented at three levels (i.e., 10 m, 46 m, and 116 m), was utilized. Previously, this tower supported the Oyster Creek Nuclear Plant operations with their emergency response procedures and now will support the decommissioning of the plant with their cleanup/disposal activities. When considering the following bullet items, the Oyster Creek (OC) tower can be considered an excellent source for the Rutgers University Center for Ocean Observation Leadership (RU-COOL) and AquaWind's sea breeze studies:

> The tower is located adjacent to the coast.
> The tower instruments/data are regulated by the Nuclear Regulatory Commission using their "stringent" QA/QC protocol (USNRC 2007).
> The tower instruments monitor winds at the heights representative of offshore wind turbine generators (WTGs). These heights are respectively stated in meters and feet in the preceding paragraph and in the following figure.

The below figures, respectively, depict the location of the Oyster Creek nuclear power plant where the met tower is installed and a diagram of the tower instrumentation:

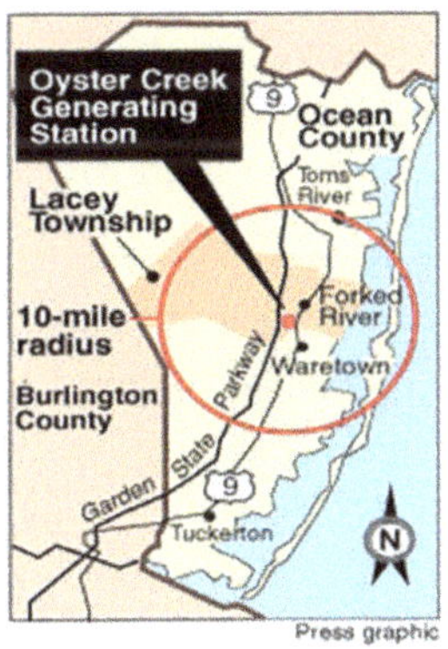

Fig. 23. OC nuclear plant/met tower location

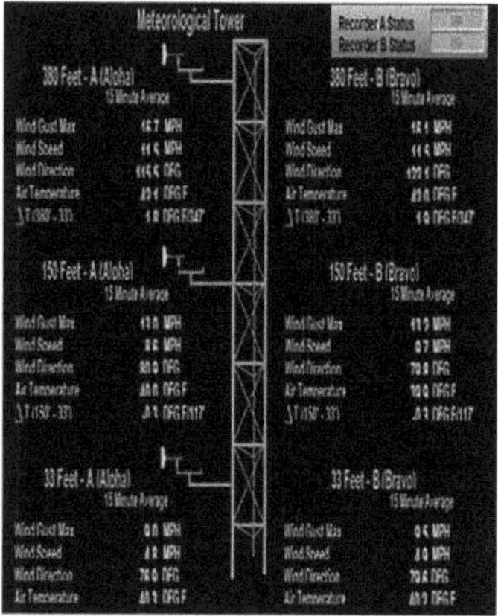

Fig. 24. OC met tower instrumentation parameters

4.2 Analytical Procedure

The data set used for this analysis includes five random years of 116 m (380 ft.) 15-minute data averaged for an hourly period, which are compiled from the 2008–2018 OC meteorological tower database. The selected data set, along with an empirical evaluation of previous and prevailing synoptic conditions (e.g., wind patterns; sky cover; and the proximity, intensity, and path of pressure systems), are utilized for the NJ sea breeze climatology analysis. This empirical evaluation, along with the observed concurrent thermodynamic properties of the local terrestrial/marine environments, is examined to identify sea breeze occurrences. Additionally, if the sea breeze develops, does it continuously continue over a significant time frame (e.g., ≥ 4 hours)? This time frame is subjectively determined to account for any significant impact the sea breeze will have on the offshore wind resource.

Since the sea breeze is a local onshore wind flow event, this analysis will focus on the climatology of sea breeze durations and intensities. Therefore, additional climatological parameters (e.g., temperature, precipitation, etc.) should be considered when analyzing the overall coastal climatology. The "preliminary" sea breeze identification criteria stated below and the subsequent analysis should provide adequate information needed for determining the "short-term" east sea breeze climatology for the central US east coast. Furthermore, the proceeding analysis is indicative of the conditions for one coastal site and not necessarily representative of other coastal locations, offshore areas, or areas located farther inland (Bowers 2004).

The following criteria for identifying onshore flow (sea breeze) occurrences are not all inclusive, and other pertinent parameters may be necessary for determining/predicting the "type" of sea breeze (i.e., *pure, backdoor, corkscrew, and synoptic*) to ascertain whether or not the onshore flow is actually a sea breeze.

> Using the Oyster Creek met tower and Tuckerton SoDAR monitoring results, select the consecutive hours when winds shift into the 32°–212° (compass degree) sector. This sector is selected to be representative of onshore flow as defined by the

central east coast shoreline configuration (Seroka 2018).

➤ Assuming both synoptic and local conditions are conducive to sea breeze development, only use the preceding wind shift criterion if the *initial* shift occurs between 8:00 a.m. and 4:00 p.m. During this time period, it should be evident that incoming solar radiation is strong enough to produce sea-to-land temperature gradients of the magnitude necessary for sea breeze development. Although favorable conditions may be apparent, there is generally a "lag" time until the actual sea breeze occurs.

➤ Depending on the land/sea/atmospheric conditions, the sea breeze can develop anytime from mid-morning to late afternoon and can continue well into the evening hours. Therefore, the hours between 1000 (10:00 a.m. ET) and 2200 (10:00 p.m. ET) can be considered the most probable time frame for sea breeze occurrences.

➤ Winds should remain continuous in the 32°–212° sector for four (4) consecutive hours or more to ensure that the shift into this sector is reasonably consistent and not the result of instrument malfunction and that the sea breeze duration is long enough to have a significant impact on coastal/offshore activities and applications.

> ➤ The resultant climatological data can then be visually displayed using suitable graphics routine(s).

4.3 Results

It should be noted that the results are expressed as *expected (average) values*. Therefore, the actual sea breeze durations and associated intensities could be significantly greater/less than the *expected values* stated in this analysis. Wind intensities are presented in both English units (mph) and metric units (m/s) with the metric unit being the measurement unit mostly used by the wind energy industry. The resultant data for this analysis are summarized in the following tables:

Table 1. Expected Sea Breeze Durations and Probability of Occurrence on Any Day of the Month

	Apr	May	Jun	Jul	Aug	Sep	Average
Hours/Day:	5.1	4.	4.	6.9	5.5	4.1	5.2
Events/Month:	15	16	14	19	15	12	15

Table 2. Expected Sea Breeze Intensities

Wind Speed	Apr	May	Jun	Jul	Aug	Sep	Average
mph	18	16	15	15	14	13	15 mph
m/s	8.0	7.2	6.7	6.7	6.3	5.8	6.8 m/s

The resulting information gained from this study can be an important asset for entities involved with the coastal/offshore environment. The most significant analytical results are listed in the following statements:

> During the sea breeze season (April–September), the number of sea breeze events along the US central east coast with their associated intensities is relatively consistent from year to year.

> When considering the concurrent adjacent land, sea, and atmospheric conditions, the resultant analytical data implies that the central US east coast sea breeze has a ~50% probability of occurring during any given day of the sea breeze season.

> Diurnal durations of a sea event have a > 20% probability of persisting during the daily 24-hour time frame. Therefore, if a sea breeze develops, it is likely that its diurnal duration will be > 4 hours.

> It is indicated that the highest probability (~60%) for sea breeze development and continued diurnal duration (~7 hours) occurs during July, with the lowest probability (~40%) and diurnal duration (~4 hours) occurring during the transitional month of September.

> Average sea breeze intensities can be very variable. This variability is primarily depen-

dent on the sea breeze "type," with the *synoptic* and *backdoor* sea breezes, respectively, exhibiting the highest average (20 mph, 8.9 m/s) and lowest average wind speed intensities (12 mph, 5.4 m/s). Average wind speed intensities for the *corkscrew* and *pure* sea breeze types are estimated to be respectively 17 mph (7.6 m/s) and 13 mph (5.8 m/s).

After analyzing the results, it becomes apparent that the *corkscrew* sea breeze is associated with relatively high average wind speeds ranging from 16 mph (7.2 m/s) to 18 mph (8.0 m/s). These higher wind intensities can be attributed to concurrent coastal upwelling events, which create a strong sea-to-land temperature gradient caused by the enhanced cooling of the sea surface as compared to the much warmer adjacent land areas. Although the *corkscrew* sea breeze is typically more intense (the *synoptic* sea breeze being the exception), its vertical and horizontal extents are normally less than most other sea breeze occurrences (Seroka 2018). The stated sea breeze climatological results are graphically summarized in the following charts:

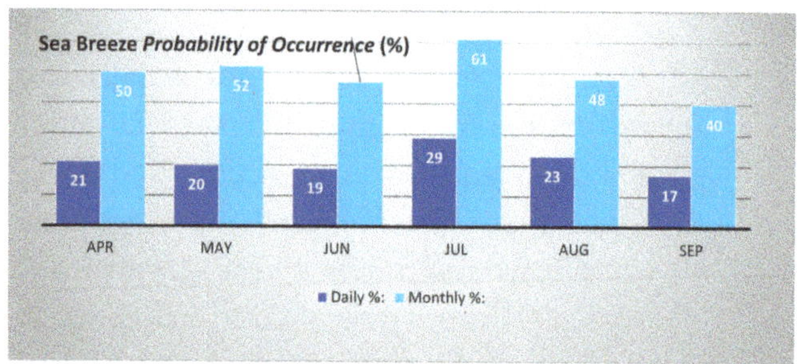

Fig. 25. Sea breeze probability of occurrence

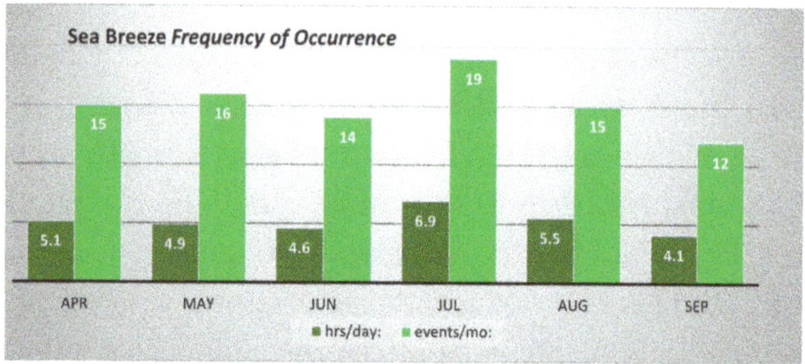

Fig. 26. Sea breeze frequency of occurrence

The results associated with the preceding charts are summarized in the below bullet points:

- Diurnal durations of a sea event have a > 20% probability of persisting during the daily 24-hour time frame. Therefore, if a sea breeze develops, it is likely that its diurnal duration will be > 4 hours.
- It is indicated that the highest probability (~60%) for sea breeze development and continued diurnal duration (~7 hours) occurs during July, with the lowest prob-

ability (~40%) and diurnal duration (~4 hours) occurring during the transitional month of September.

➢ Average sea breeze intensities can be very variable. This variability is primarily dependent on the sea breeze "type," with the *synoptic* and *backdoor* sea breezes, respectively, exhibiting the highest average wind speeds (20 mph, 8.9 m/s). Average wind speed intensities for the *corkscrew* and *pure* sea breeze types were estimated to be respectively 17 mph (7.6 m/s) and 13 mph (5.8 m/s).

To determine whether or not observed onshore flow is a "true" sea breeze, along with estimating *extreme values* (max/min), *expected values* (averages), and the *variability* associated with the sea breeze circulation, additional advanced statistical procedures (e.g., *discriminant/extreme value analyses*) should be utilized. Also, to ensure that the data is statistically significant, a more detailed/extended data set (e.g., a period of record $\geq$ 10 years) in conjunction with shorter averaging times (e.g., 1 to 10 minutes) rather than the 1-hour averaging time currently utilized could improve the reliability and representativeness of the resultant climatological analysis. Note that 10-min averaging times are considered the "standard" for the wind energy industry. When considering the preceding statements, more realistic parameter values that "best" define the sea breeze climatology can be achieved.

5. Monitoring and Modeling Methodology Summary

Since atmospheric monitoring by most in-situ and remote systems does not cover the spatial extent for the onshore and offshore components of the sea breeze circulation, using monitoring in conjunction with representative model simulations (e.g., RU-WRF + LCS technology) along with LES technology, in addition to a potentially coupled ocean/atmosphere model, is probably the "best" methodology for determining the spatial and temporal variability of the sea breeze circulation (Seroka 2018). This proposed methodology, when used in conjunction with the VMT concept and applying the most advanced climatology data, should provide the most accurate and representative diagnostics/predictions for sea breeze circulation.

6. Renewable Energy Applications

As was implied in prior sections, the sea breeze is a local wind circulation, and its most significant effect will be associated with *coastal/offshore wind energy* design, development, and implementation. Examples of coastal/offshore wind energy facilities are shown in the respective pictures:

Fig. 27. Coastal wind energy facility, Atlantic City, NJ

Fig. 28. Offshore wind energy facility, Block Island, RI

It should be mentioned that the offshore component of the lower portion of the sea breeze cell will affect wave heights, frequencies, and intensities. These factors will either enhance or minimize *wave energy* production, as they are dependent on the speed and direction of the sea breeze flow in relation to the location of the wave energy facility. Also, when analyzing the potential convective characteristics of the sea breeze circulation, especially near the sea breeze "front", cloud formation and possible storm development could occur, causing reduced incoming solar radiation, which will negatively impact *solar energy* production for facilities located in these areas of convective activity.

The most recent sea breeze analyses related to coastal/offshore wind energy endeavors reveal that 3D sea breeze characteristics (e.g., wind vector gradients/profiles, turbulence intensity, and shear factors) will have a substantial effect on wind turbine generator (WTG) site location, design specifications, array configurations, and resultant power production (Dunk et al. 2015). Considering the effects of sea breeze circulation, the following factors provide an insight into coastal/offshore wind energy power production for a substantial portion of the US east coast:

> Wind speeds are estimated to range from ~6 m/s nearer the coast and decrease to ~2 m/s farther offshore as a result of intense sea breeze development. This estimated wind speed gradient is contrary to most

offshore wind resource assessments, which assume wind speeds increase from the coast to offshore areas. However, modeling simulations show wind speeds can increase from ~6 m/s closer to the coast to near 11 m/s farther offshore over areas where the sea breeze did not develop.

➤ The sea breeze circulation is considered a dynamic process, with wind flow vectors that constantly change over the diurnal cycle of the circulation. Depending on the individual WTG design, most offshore WTGs will *not* start to produce power until wind speeds exceed ~3 m/s to 4 m/s (i.e., *self-start or cut-in speed*).

The preceding bullet points indicate that WTGs potentially located within certain offshore areas can be producing power, while other WTGs will be idle during a specific sea breeze event. As the sea breeze propagates farther offshore and inland, WTGs that were idle can start to produce power, and WTGs that are producing power may become idle. Consequently, the magnitude of dynamic interactions at the land/sea/air interface will determine the range of values for sea breeze durations and intensities. Modeling applications would then effectively support electrical grid management, the development of power transportation, and distribution procedures that will provide the most efficient "mix" of renewable, fossil fuel, and nuclear generation needed to accommodate the energy demand.

The RU-WRF model results are evaluated to estimate potential power production for "hypothetical" offshore wind facilities located along the central east coast.

Lease zone delineations off the coast of NJ, along with representative offshore WTG dimensions, are respectively shown in the following figures. (Note: larger WTGs with generating capacities ranging from 8 MW to 12 MW will probably be used for forthcoming offshore wind energy applications.)

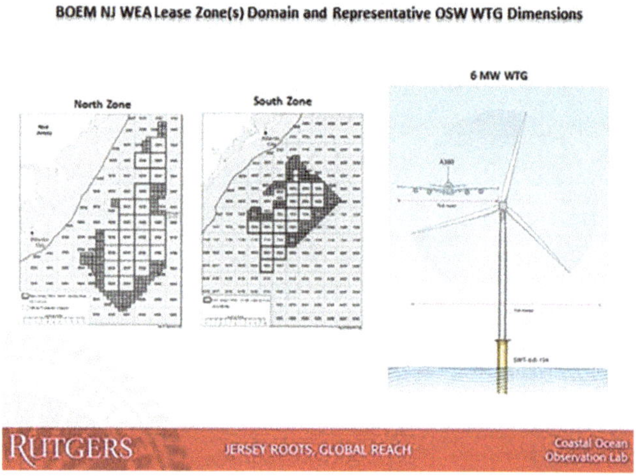

Fig. 29. NJ offshore wind energy lease zones and representative offshore wind energy turbine dimensions

An additional analysis of the variability of the gross capacity factor (GCF) resulting from the sea breeze circulation was conducted using the following procedures. The RU-WRF model was run at a 3-kilometer grid resolution for the *internal* modeling domain, which encompasses offshore/inland areas associated with the central east coast and extends ~100 miles offshore. This analysis indicates the GCF can range from ~20% to > 80%, which is very significant when considering that the cost benefit of offshore wind power production is dependent on the amount of available energy (supply) and consumer requirements (demand), especially during times of "peak" energy demand (PJM 2019). The stated range of GCFs is "nested" to run at microscale resolutions (< 2-kilometer grid spacing) to resolve the local wind circulations (e.g., the sea breeze) that cause perturbations within the general flow patterns of the offshore wind resource. The *external* modeling domain resolution, which was used to set initial boundary conditions for the higher-resolution model runs, was set for a 9-kilometer grid spacing.

Simulations of the selected grid spacing options are depicted in the following simulations:

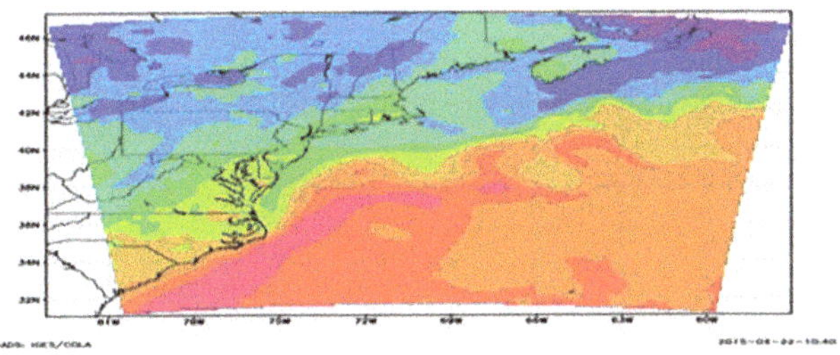

9-kilometer grid resolution mesoscale *external* modeling domain

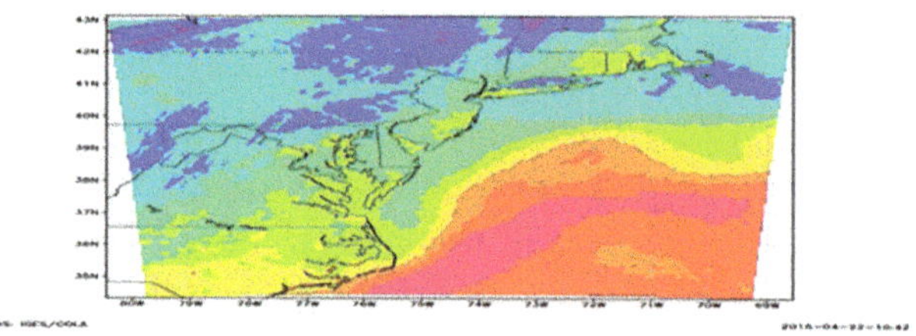

3-kilometer grid resolution mesoscale *internal* modeling domain

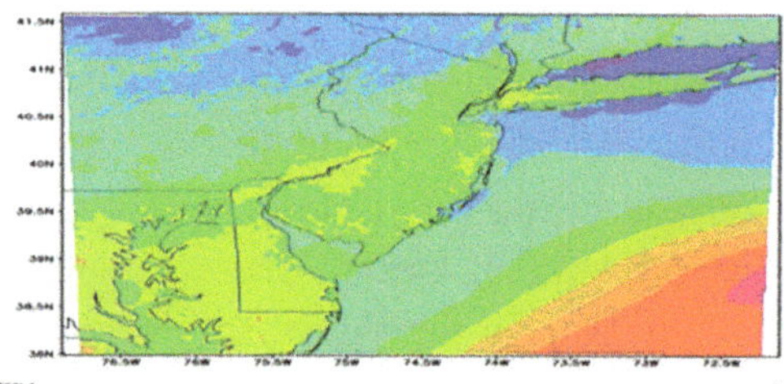

0.4 < 2 km grid resolution microscale *local* modeling domain

Figs. 30–32. Gross capacity factor estimations using RU-WRF modeling domains ranging respectively from 9-kilometer grid resolution to near 0.2-kilometer grid resolution needed to realistically represent the GCF resulting from offshore wind power production

Using the preceding RU-WRF modeling grid resolutions, an analysis of the average diurnal variation of offshore wind speeds compiled at 100 m over the central east coast, along with the average daily energy demand, is displayed in the respective graphs shown below:

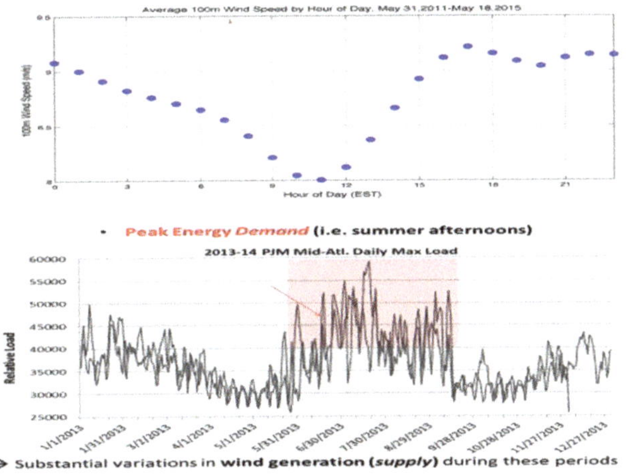

Figs. 33–34. Peak energy demand periods and concurrent potential offshore wind power production

The preceding analysis of average diurnal variation in offshore winds when compared to the variation in energy demand provides the following evidence regarding generation load requirements:

> Adequate offshore wind facility capacity located within an area with a significant wind resource at the suggested 100-meter hub height should significantly supplement conventional generation load requirements. This assumption is especially applicable during periods of "peak" energy demand and will, therefore, reduce the burden put on the available energy supply and grid transmission/distribution during these critical times. Average hourly wind

speeds at 100 m above MSL that coincide with the preceding statements are summarized as follows:

- The average hourly offshore wind speed at 100 m above MSL was lowest at 11:00 a.m. (~8.0 m/s) and highest at 5:00 p.m. (~9.2 m/s).
- Wind speeds remain high until 5:00 p.m. to 11:00 p.m. (9.1 m/s), and then slowly drop off to 8.0 m/s by 11:00 a.m. the next day.

➢ To reduce the "uncertainty" associated with wind variability, local wind perturbations frequently caused by sea breeze circulation should be accounted for to determine when and where offshore wind power production will be available during the summer season when the "primary peak" energy demand period occurs.

➢ Coastal storms that impact the offshore wind resource during the winter "secondary peak" energy demand period should be taken into account to determine the availability of offshore wind power production.

The below table provides a summary of the central east coast wind resource:

Table 3. Average Coastal/Offshore Wind Speeds

Average Wind Speed (WdSp) at 100 m	Avg Max WdSp.	Avg WdSp	Energy Demand Period	Normal	Avg Min WdSp	Avg WdSp	Energy Demand Period	Normal
	10.5 m/s	9.5 m/s	Winter	10.63 m/s	5.7 m/s	7.0 m/s	Summer	7.23 m/s

The implications regarding the offshore wind resource over the central east coast at 100 m above MSL are summarized below:

> ➢ The current average wind speeds, which are estimated to occur during the winter "secondary peak" energy demand period, appear to be below *normal*; also, average wind speeds estimated to occur during the summer "primary peak" energy demand period are also below *normal*.

> ➢ The findings stated on the previous page are somewhat contrary to "popular" climate change/global warming theories that suggest that wind and storm intensities are increasing. Overall, this may prove to be true; however, possibly the mesoscale sea breeze circulation can exhibit its own unique characteristics as being dependent on the dynamics and physical properties of the local land/sea/air interactions that

occur during sea breeze development and intensification.

➤ Wind speeds and diurnal durations of frequent sea breeze occurrences observed are not, respectively, intense enough or long enough to be accounted for in the "long-term" (i.e., seasonal and annual) averages. This suggests that sea breeze occurrences should be evaluated on a case-by-case basis.

➤ It is implied that higher modeling resolutions, along with representative initial boundary conditions, including physics input that is realistic for the offshore environment, should resolve the dimensional and flow properties of sea breeze circulation, especially for the characteristics related to the offshore component of sea breeze circulation.

➤ "Short-term" (i.e., diurnal and monthly) variability should be accounted for in the planning and design process for coastal/offshore facilities to reduce the "risks" related to abnormal impacts that could be caused by the dynamics of sea breeze circulation.

The implications involved with the relationship between wind speed and power production indicate that an increase (decrease) of only ~1.0 m/s in WTG hub-height wind speed can be significant for potential power production [i.e., wind power production is proportional to the cube of the wind speed (m/

s³)]. Coastal storms can cause increased wind speeds, resulting in an increase in coastal/offshore wind power production. Although sea breeze wind speeds are not as intensive as those produced by coastal storms, they are more consistent and will, therefore, have a more significant effect on power production. The effects of the sea breeze on wind power production will be similar to those experienced by coastal storms but will be less intense, occur more often, especially during mid-spring through early fall, and, because of its variability, may have to be evaluated on a case-by-case basis. Furthermore, when compared to coastal storms, the sea breeze will be much less likely to cause power outages and damage to the electrical grid infrastructure.

Depending on the WTG design specifications, self-start-up ("cut-in") speeds will occur when wind speeds are ~ 3 to 4 m/s. When wind speeds are at or slightly above the start-up speed, power generation and total energy produced along with GCFs will be minimal. When winds are higher than the WTG automatic shut-down ("cut-out") speed (i.e., ~25 to 30 m/s depending on WTG design specifications), the WTGs will terminate operation and, therefore, power generation and total energy produced along with GCFs will be zero. An example of the variability of GCFs associated with sea breeze circulation is presented in the proceeding simulation:

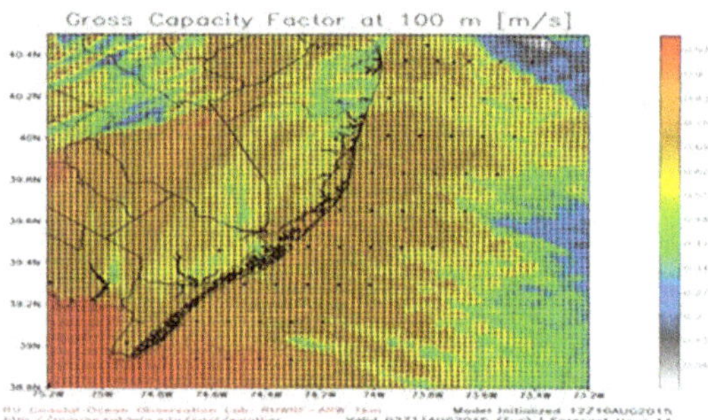

Fig. 35. A "snapshot" of offshore WTG GCFs at 100 m estimated for a sea breeze occurrence over coastal/offshore waters. The GCFs ranged from ~20 to > 80%

7. Environmental Assessments

As discussed in prior sections, sea breeze circulation has both positive and adverse effects on the economy, energy production/transmission and distribution, health, and environmental resources. Regarding environmental assessments, the following issues should be analyzed and implemented to ensure that the information associated with sea breeze circulation is utilized in the most cost-effective and efficient manner to ensure that our environmental resources are maintained and, hopefully, improved so they will enhance the quality of our coastal/offshore environment.

The sea breeze's unique circulation capabilities will not only transport and disperse gaseous, particulate, and odorous matter but will also recirculate these contaminants, resulting in a cumulative effect of adverse air pollution concentrations. These adverse conditions are primarily caused by emissions from sources located within or adjacent to coastal areas. These sources include mobile (e.g., cars, trucks, buses, and aircraft flying into and out of local airports) and stationary facilities (e.g., power plants and industrial units).

The recirculation process also applies to "natural" emissions caused by wildfires that occur in forested areas located in coastal areas, along with certain health-related matter, which can be attributed not only to detrimental air contaminants but also to the effects of pollen and/or other allergens that are transported and dispersed by the sea breeze circulation (Bielory 2021). Furthermore, the recirculation of these contaminants offshore can cause deposition in seawater, resulting in increased adverse water quality. Although sea breeze circulation can cause adverse environmental and health impacts, it can also be beneficial to areas that are basically void of detrimental airborne substances. Over these areas, the sea breeze can provide a cooling and "cleansing" effect produced by the relatively "clean" onshore maritime air, which could potentially improve health problems and mitigate environmental issues.

The structure of the sea breeze cell that can produce the previously stated adverse environmental conditions is depicted in the following diagram, which is self-explanatory [note: although the term "lake" is used for the offshore water area, this schematic applies to most land/water/air interactions that occur near large bodies of water (i.e., oceans)]. LCL refers to the lifting condensation level where, possibly, clouds and storm development can occur. The TIBL is the "thermal internal boundary layer," which is formed by the cooler onshore flow interacting with the warmer terrestrial air. The TIBL then becomes the upper boundary or "mixing height" for air pollutants contained within this lower air layer, as defined in the below diagram (Petersen 1999):

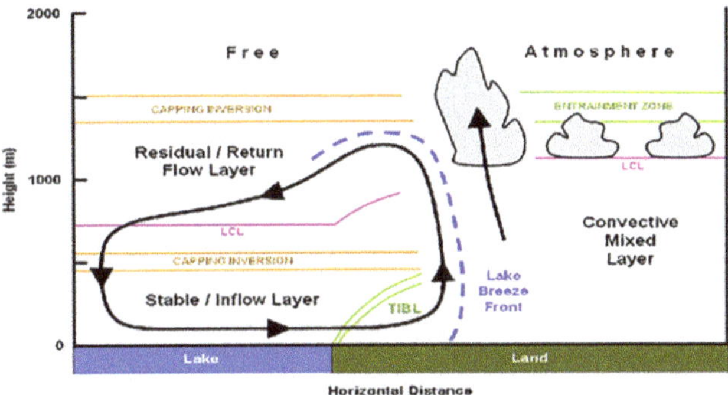

Fig. 36. Sea breeze circulation structure showing both the onshore/offshore components of the sea breeze cell

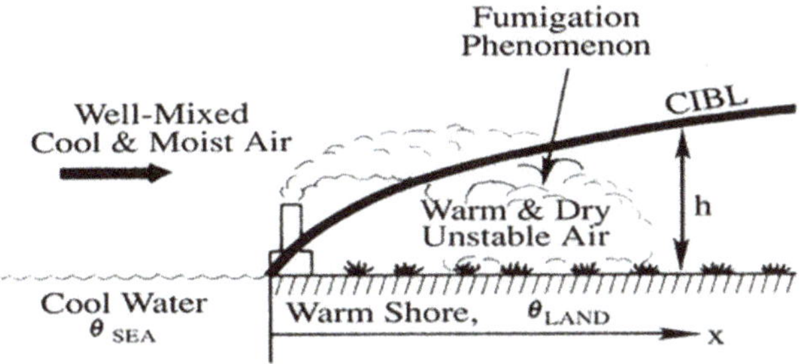

Fig. 37. A depiction of the convective thermal internal boundary layer (TIBL) that develops during the sea breeze initiation and intensification as the sea breeze propagates over inland areas. Trapping, fumigation, and recirculation of air contaminants can occur over time while the sea breeze flow is sustained.

The recirculation process, potential "fumigating" conditions of air contaminants plus other airborne substances "trapped" within the TIBL, and its effect on both the onshore and offshore environment are shown in the above schematic (note: Referring to Fig. 37, the CIBL is the convective internal boundary layer, which is the same as the TIBL; "h" denotes the mixing height as measured from the surface to the base of the CIBL):

Following are illustrations of, respectively, the horizontal and vertical perspectives of upper-level and lower-level emission plumes affected by the sea breeze circulation. The white upper-level plume is transported offshore by the general synoptic flow, while the white lower-level plumes are transported inland by the sea breeze circulation.

Fig. 38. Horizontal perspective of emission plume transport above and within the sea breeze circulation

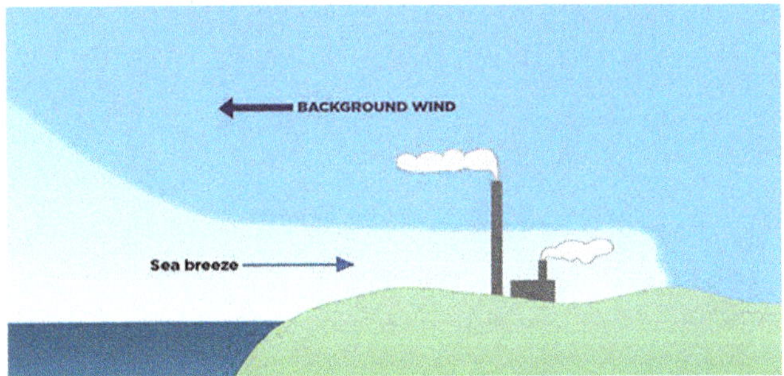

Fig. 39. Vertical perspective of emission plume transport above and within the sea breeze circulation

8. Recreational Activities

Several recreational activities associated with coastal areas are affected by the sea breeze circulation. A description of the dynamics that cause the sea breeze to develop and be sustained is presented in previous sections of this publication. The recreational endeavors affected by the sea breeze would include, but not limited to, the following activities:

❖ **Airborne Activities:**
  ➢ Small aircraft sport flying (e.g., prop propelled aircraft including ultra-lights; gliders (soaring)).
  ➢ Hot air ballooning.
  ➢ Parasailing; hang gliding.
  ➢ Kite flying.

❖ **Surface-Based Activities:**
  ➢ Sport Boating (motor boating/sailing); boat racing.
  ➢ Offshore, bay, and surf fishing.
  ➢ Wind surfing; conventional surfing; body boarding.
  ➢ Water skiing; kite/wake boarding, tubing.
  ➢ Kayaking; paddle boarding.
  ➢ Ocean bathing, swimming, and snorkeling.
  ➢ Beach sports (e.g., volleyball).
  ➢ Beach sunbathing.

➢ Boardwalk shopping/amusements.

The preceding recreational activities are depicted in the following images:

❖ **Airborne Activities:**

Fig 40. Private Single Engine Aircraft

Fig 41. Ultra-Light Aircraft

Fig 42. Soaring (glider) Aircraft

Fig 43. Hot Air Ballooning

Fig 44. Parasailing

Fig 45. Hang Gliding

Fig 46. Kite Flying

❖ **Surface-Based Activities:**

Fig 47. Motor Boating

Fig 48. Sailing

Fig 49. Boat Racing

Fig 50. Offshore Fishing

Fig 51. Bay Fishing

Fig 52. Surf Fishing

Fig 53. Wind Surfing

Fig 54. Conventional Surfing

Fig 55. Body Boarding

Fig 56. Water Skiing

Fig 57. Kite Boarding

Fig 58. Wake Boarding

Fig 59. Tubing

Fig 60. Kayaking

Fig 61. Paddle Boarding

Fig 62. Ocean bathing/swimming

Fig 63. Snorkeling

Fig 64. Beach Sports

Fig 65. Sunbathing

Fig 66. Seaside Shopping/Amusements

After reviewing the preceding images related to the activities that could be affected by the sea breeze, a working knowledge of sea breeze dynamics and its realistic prediction will enhance the cost-effectiveness of the planning process for these events. This process will ensure that the resultant enjoyment and cost-effectiveness of applications associated with the several recreational activities relevant to coastal areas will be realized. In many coastal areas the sea breeze occurs almost daily from the spring through early fall seasons. As previously discussed, this onshore air flow is caused by the temperature differential between the cooler sea surface and adjacent warmer Land. The sea breeze usually begins from late morning to early afternoon reaching a maximum during the mid to late afternoon and then subsides after the land has cooled during early to late evening.

The leading edge, or "front," of the propagating cool sea breeze forces the inland warmer air to rise. This forced convection process can create strong updrafts along with possible thunderstorm development. Consequently, the airborne activities could experience adverse impacts if they are conducted in the atmospheric layer associated with this area of wind shear and convective activity. Furthermore, because of gaseous, liquid, and solid substances being transported upward along the front, visibility may decrease. The converse could occur in the area associated with the offshore sea breeze front where downdrafts caused by subsiding air will produce less intense winds with relatively clear sky conditions. If airborne activities occur within the sea breeze circulation, an increase in persistent wind speeds will be observed when compared to the external air flow.

Airborne activities that are mobile can approach and interact with the sea breeze circulation and, therefore, experience various changes in vertical and horizontal wind shear, turbulence, temperature, and visibility conditions. For example, a small plane pilot operating near coastal areas often will experience enhanced lift, wind shear, turbulence and deceased visibility at the altitude mostly associated with the advecting inland sea breeze front. Sea breeze flow vectors described in the preceding statements are depicted in the following diagram:

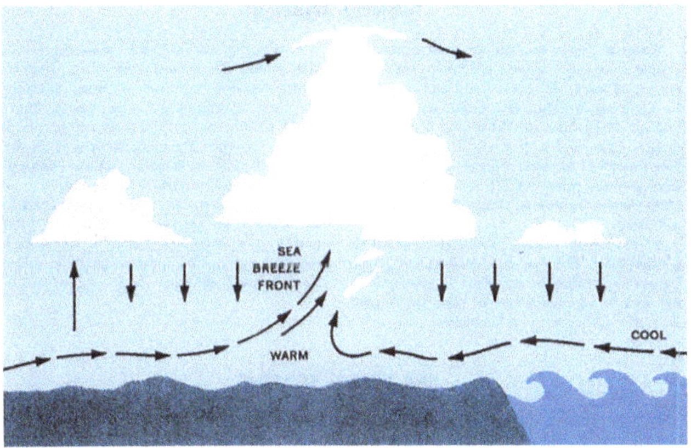

Fig 67. Schematic of cross section of the sea breeze circulation showing horizontal and vertical flow vectors. If the air inland is moist, cumulus cloud development often marks the sea breeze front.

Regarding surface-based activities, the sea breeze can have various impacts on surface currents/waves, wind intensities, and temperatures, These impacts will not only be experienced in coastal waters but will also occur over inland locations, such as adjourning bays and tributaries. Understanding the sea breeze and when to expect its occurrence can enhance the quality of recreational activities on or near the water. Rather than just reviewing the general aspects of the local weather forecast, the forecast along with an individual's experience should be analyzed to ascertain when a sea breeze will occur and take advantage of this information to ensure that planned recreational activities will be safe and enjoyable. This may mean that the activity could be postponed or changed to a different location and/or time to ensure the activity is conducted in a safe and positive manner.

Depending on the pending activity or activities, the effects of the sea breeze on these activities could either be positive or negative as determined by the nature of the activity and the criteria selected to ensure the activity will satisfy the objectives of the anticipated event. As winds blow toward the shore, they create small waves known as wind "chop." This wind-chop produces a disjointed wave, with small choppy waves disrupting the naturally occurring larger swells (waves). Winds blowing offshore affect waves differently because they don' have a large fetch (i.e., wind fetch is the uninterrupted distance over which wind blows across a body of water). Therefore, offshore winds do not create any chop in the water, instead they have a grooming effect, which will smooth out any chops and "wrinkles" in the waves, resulting in a clean, uniform wave formation.

Considering all aspects of the sea breeze, the objectives of most surface-based activities will be best realized during non-sea breeze events because surface water and near surface winds conditions are more uniform, stable, and predictable. The pressure (temperature) and flow patterns associated with sea breeze development are shown in the following respective diagrams.

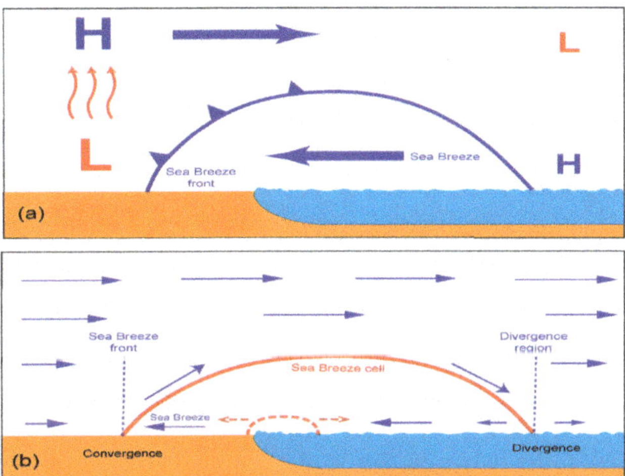

Fig 68. (a) Pressure (temperature) differences relevant to land/sea surface conditions and the overlying atmosphere (blue designates colder temperatures/higher pressures; red designates warmer temperatures/lower pressures. (b) Wind flow vectors relative to the sea breeze circulation; convergence coincides with warmer rising air and divergence coincides with cooler subsiding air.

Another aspect of the sea breeze and its interaction with near-shore waters is the effects of coastal upwelling, which was discussed in previous sections of this publication. When strong southerly winds cause warmer sea surface waters to be replaced with underlying colder waters, local sea breezes could intensify resulting from the extreme temperature difference between the resultant colder sea surface temperatures and warmer land temperatures. The increase in along shore or onshore wind flow intensity, in addition to the dramatic decrease in sea surface and adjacent inland air temperatures, could cause a reduction in beach activities with an increase in boardwalk or near shore entertainment endeavors. These endeavors could include activities such as shopping and participation in the various amusements that are available in these coastal locations.

In summary, the sea breeze has a definite impact on both airborne and surface-based recreational activities that are associated with the coastal environment. Whether this impact is negative or positive will be determined by the specific situation that is affected by the sea breeze and what is expected by those involved to ensure a pleasant and rewarding experience.

9. Concluding Remarks

Representative monitoring systems, along with verified high-resolution modeling programs, including the possibility of the application of a "coupled" atmospheric/ocean model, should be considered to produce more accurate estimates of the land/sea surface conditions and concurrent atmospheric boundary layer characteristics that control sea breeze development and intensification. Additionally, model results could be utilized to support various activities associated with energy/engineering, economic, and environmental assessments to provide a more automated, efficient, and reliable data exchange process. The following diagnostic/predictive parameters should be investigated to determine their benefits in providing a better understanding of sea breeze structures and associated sea breeze dynamics (Dunk et al. 2015):

> RU-WRF + LCS/RD model simulations:

- Horizontal temperature/wind speed/pressure/moisture gradients associated with the land/coastal ocean interface
- Vertical temperature/wind speed/pressure/moisture profiles at the land/coastal ocean interface, along with areas farther inland and offshore

> ➤ Prior and prevailing synoptic parameters at an atmospheric level (e.g., 925 mb) most representative of the sea breeze circulation

Although most of the information presented in this publication is based on studies conducted for NJ and adjacent coastal areas, the results and concepts provided should be applicable for most coastal areas with similar topography, shoreline configurations, and atmospheric synoptic flow characteristics. Consequently, the suggested current and proposed monitoring/modeling programs can then be utilized to support the following applications:

> ➤ Prediction and analysis of sea breeze and non-sea breeze occurrences, along with sea breeze type
> ➤ Prediction and analysis of upwelling and non-upwelling events
> ➤ Determination of atmospheric transport/ dispersion variables and how they impact both air and water quality

There are several applications in this writing that are relevant to coastal areas affected by sea breeze circulation. These applications could range from recreational to commercial endeavors related to fisheries management, energy production/transportation/ distribution, environmental management, homeland security, health issues, aviation activities (especially wind shear impacts during take offs/landings), and economic growth/stabilization.

Therefore, a working knowledge of the sea breeze and its effects on the aforementioned applications should be a significant portion of any design, planning, and operational procedures associated with coastal/offshore areas. Therefore, whether the effects of the sea breeze are positive or negative, they should be accounted for to ensure that any activities conducted in the coastal/offshore environment are efficient and cost-effective.

Acknowledgments

The professional contributions to the manuscript preparation and associated figures/graphs can be attributed, respectively, to my former meteorology graduate students: Lou Bowers, Greg Seroka, and Erick Fredj, a visiting professor of computational studies at Rutgers University.

References

Avissar, R., Moran, M. D., Wu, G., Meroney, R. N., and R. A. Pielke. 1990. "Operating Ranges of Mesoscale Numerical Models and Meteorological Wind Tunnels for the Simulation of Sea and Land Breezes." *Boundary Layer Meteorology* vol. 50 (1990): 227–275.

Bielory, L., Bowers, L., Marcus, R., and R. Dunk. 2021. "The Influence of Sea Breeze on Mold Dispersion." *Allergy and Asthma Proceedings* 42, no. 3: 222–227.

Bodini, N., Hu, W., Optis, M., Cerone, G., S. Alessandrini. 2023. "Long-Term Uncertainty Qualification in WRF Modeled Offshore Wind Resource off the US Atlantic Coast." *Wind Engineering Science* 8: 607–620. https://doi.org/10.5194/wes-8-607-2023.

Bowers, L. 2004. "The Effect of Sea Surface Temperature on Sea Breeze Dynamics Along the Coast of New Jersey." MS thesis, Institute of Marine and Coastal Sciences (IMCS), Rutgers University, New Brunswick, NJ 08901.

Dunk, R., Glenn, S., Kohut, J., Miles, T., Seroka, G., and E. Fredj. 2015. "Evaluation of New Jersey's Offshore Wind Resources: Atmospheric/Oceanic Modeling and Analyses to Support the Offshore Wind Energy Development Process as Defined in New Jersey's Offshore Wind

Economic Development Act (OWEDA)." New Jersey Board of Public Utilities (NJBPU), Division of Economic Development & Energy Policy and the Office of Clean Energy, RU Grant # BPU-070, Rutgers Center for Ocean Observation Leadership (RU-COOL) Rutgers Dept. of Marine and Coastal Sciences (DMCS), New Brunswick, NJ 0890. http://rucool. marine.rutgers.edu/bpu/.

Fischereit, J., Brown, R., Gue, X., Badger, J., and G. Hawkes. 2022. "Review of Mesoscale Wind-Farm Parametrizations and Their Applications." *Boundary Layer Meteorology* 182: 175–224. https://doi.org/10.1007/s10546-021-00652-y.

Fredj, E., Carlson, D., Amitai, Y., and H. Gozolchiani. 2016. "The Particle Tracking and Analysis Toolbox (PaTATO) for Matlab." *Limnology and Oceanography Methods* vol. 14: 586–599.

Grannakopoulou, E-M. and R. Nhili. 2014. "WFR Model Methodology for Offshore Wind Energy Applications." EDF Energy R&D UK Center, London, SW1W0, AUF, 24 Mar 2014.

Grosman, E. and Horel, J. 2012. "Idealized Large-Eddy Simulation of Sea and Lake Breezes: Sensitivity to Lake Diameter, Heat Flux, and Stability." *Boundary Layer Meteorology* 144: 309–328.

Lyons, W. and L. Olsson. 1972. "Mesoscale Air Pollution Transport in the Chicago Lake Breeze." *Journal of the Air Control Association* SSN: 0002-2470.

Optis, M., Kumler, A., Scott, G., Debusth, M., and P. Moriarty. 2020. "Validation of RU-WRF, the Custom Atmospheric Mesoscale Model of the Rutgers Center for Ocean Observing Leadership." National Renewable Energy Laboratory (NREL), Strategic Partnership Project Report (FIA-18-01872), NREL/TP-5000-75209.

Peng, S., Kon, Y., and H. Watanabe. 2022. Effects of Sea Breeze on Urban Areas Using Computational Fluid Dynamics-A Case Study of the Range of Cooling and Humidity Effects in Sendai, Japan, Sustainability." *Sustainable Urban and Rural Development* 14, no. 3: 1074. https://doi.org/10.3390/su14031074.

Petersen, R. 1999. "Development and Evaluation of an Improved Algorithm for Evaluating Thermal Internal Boundary Layer Heights." *Journal of Wind Engineering and Industrial Aerodynamics* (April 1999).

PJM. 2019. "Summer Operations of the PJM Grid: June 1–Sept 15, 2019." PJM Interconnection, Valley Forge, PA, Internal Report (available for public review) October 2019.

Robinson, D. 2022. "Climatology including the Sea Breeze, Transition Complete: May and Spring 2022." *NJ.* Internal Publication, Center for Environmental Prediction, School of Environmental & Biological Sciences/ NJAES, Rutgers University, New Brunswick, NJ.

Robinson, D. and D. Ludlam. 2020. "NJ Climate Overview, Coastal Zone." Office of the NJ State Climatologist (ONJSC), Dept. of Geography, Rutgers University, Piscataway, NJ 08854. http://climate.rutgers.edu/stateclim_v1/njclimoverview.

Seroka, G., Fredj, E., Kohut, J., Dunk, R., Miles, T., and S. Glenn. 2018. "Sea Breeze Sensitivity to Coastal Upwelling and Synoptic Flow Using Lagrangian Methods." *Journal of Geophysical Research: Atmospheres* 123, no. 17: 9443–9461. https://api.sematicscholar.org/corpus.

Seroka, G. and E. Fredj. 2020. "FY20 AquaWind Lagrangian Coherent Structures (LCS) Final Report: 3D LCS Proof of Concept for Sea Breeze." Internal report submitted to the Rutgers University Center for Ocean Observation Leadership (RU-COOL) in support of the NJBPU Offshore Wind Energy Development Initiative.

Simpson, J. 2007. *Sea Breeze and Local Winds.* England: Cambridge University Press.

USNRC. 2007. "Standard Review Plan 2.3.3" On-Site Meteorological Measurements Program, NUREG-0800, US Nuclear Regulatory Commission.

About the Author

Rich Dunk, PhD, CCM is the owner and principal meteorologist/senior advisor for AquaWind, which is a limited liability company (LLC) registered as a small business in the state of New Jersey. AquaWind is a consulting company that specializes in the analyses/predictions of atmospheric dynamics associated with mesoscale and microscale processes that impact the coastal/offshore environment. Rich started his career in 1967 as an earth science teacher and then entered the US Army in 1968. He completed the Army's meteorology training program as an honor graduate. Rich then served as a weather forecaster and ballistic meteorologist in support of the artillery fire missions and helicopter deployments. His duty assignments included Arizona, Vietnam, and the atmospheric science lab located in Fort Monmouth, New Jersey. During his military career, Rich received several commendations and awards. After he was discharged from the military service in 1971, Rich was employed as an environmental specialist for Research Cottrell. In 1974, he became employed by the US Metals Refining Co., AMAX, Inc., as an environmental engineer. He then advanced from the environmental engineer position to become the senior environmental engineer and then the manager for environmental services.

After leaving AMAX in 1990, Rich was employed by Jersey Central Power & Light (JCP&L, a GPU company) as the senior environmental scientist in charge of all aspects of air quality management. While employed at JCP&L, he was selected to serve on the Electric Power Research Institute (EPRI) Task Force committees involved with improving atmospheric dispersion modeling programs and emission monitoring technology being utilized by the electric utility industry. Along with his colleagues at EPRI, his efforts in these activities resulted in a new shoreline mixing depth model, a more realistic plume rise algorithm, and a cost-effective predictive emissions monitoring procedure. Following his tenure at JCP&L, Rich was promoted to the corporate level as a staff meteorologist to support the GPU Power Services group, which was responsible for energy procurement, selling, and trading activities. He retired from GPU in 2002 and began to work at Rutgers as an adjunct professor and research scientist for, respectively, the Environmental Sciences Department, Rutgers University Center for Environmental Prediction (RUCEP), and the Rutgers University Coastal Ocean Observation Laboratory (RU-COOL), which is now the Center for Ocean Observation Leadership.

Rich currently specializes in offshore wind resource assessments. These assessments include studies of area-specific wind patterns, such as sea breeze circulations and coastal storms, which impact the wind resource and resultant flow characteristics. These applied studies are conducted to support cost-effective offshore wind energy development, improve coastal storm forecasting, and enhance the efficient planning and operations of the utility industry. He received a BA in science education (1967) from Glassboro State College (now Rowan University), a BS in environmental sciences/meteorology (1972) with the highest honors from Rutgers University, where he was inducted into the Meteorology Honor Society (Chi Epsilon Pi), an MS in meteorology (1976) from Rutgers University, and a PhD (1994) in environmental science/atmospheric science from Rutgers University. Dr. Dunk was board-certified as a consulting meteorologist in 1979 by the American Meteorological Society (AMS) and has retained the status of a certified consulting meteorologist (CCM) through the current time.